W0018726

Topological Dynamics in Metamodel Discovery with Artificial Intelligence

The leveraging of artificial intelligence (AI) for model discovery in dynamical systems is cross-fertilizing and revolutionizing both disciplines, heralding a new era of data-driven science. This book is placed at the forefront of this endeavor, taking model discovery to the next level.

Dealing with artificial intelligence, this book delineates AI's role in model discovery for dynamical systems. With the implementation of topological methods to construct metamodels, it engages with levels of complexity and multiscale hierarchies hitherto considered off limits for data science.

Key Features:

- Introduces new and advanced methods of model discovery for time series data using artificial intelligence

- Implements topological approaches to distill "machine-intuitive" models from complex dynamics data

- Introduces a new paradigm for a parsimonious model of a dynamical system without resorting to differential equations

- Heralds a new era in data-driven science and engineering based on the operational concept of "computational intuition"

Intended for graduate students, researchers, and practitioners interested in dynamical systems empowered by AI or machine learning and in their biological, engineering, and biomedical applications, this book will represent a significant educational resource for people engaged in AI-related cross-disciplinary projects.

Chapman & Hall/CRC
Artificial Intelligence and Robotics Series
Series Editor: Roman Yampolskiy

For more information about this series please visit:
https://www.routledge.com/Chapman--HallCRC-Artificial-Intelligence-and-Robotics-Series/book-series/ARTILRO

Topological Dynamics in Metamodel Discovery with Artificial Intelligence

From Biomedical to Cosmological Technologies

Ariel Fernández

CRC Press
Taylor & Francis Group
Boca Raton London New York

CRC Press is an imprint of the
Taylor & Francis Group, an **informa** business
A CHAPMAN & HALL BOOK

Access the Supplemental Material: https://www.routledge.com/9781032366326

First Edition published 2023
by CRC Press
6000 Broken Sound Parkway NW, Suite 300, Boca Raton, FL 33487-2742

and by CRC Press
4 Park Square, Milton Park, Abingdon, Oxon, OX14 4RN

CRC Press is an imprint of Taylor & Francis Group, LLC

© 2023 Ariel Fernández

ISBN: 978-1-032-36632-6 (hbk)
ISBN: 978-1-032-36633-3 (pbk)
ISBN: 978-1-003-33301-2 (ebk)

DOI: 10.1201/9781003333012

Typeset in Minion
by SPi Technologies India Pvt Ltd (Straive)

In loving memory of Haydée Stigliano, my mother

Contents

Preface

In his proposal for the construction of a digital mind, Ray Kurzweil naturally starts by addressing the problem of how the mind works, a sobering task, as Francis Crick would have it. After a long cogitation, Kurzweil concludes that a connected hierarchy of pattern recognizers is essential and that contents or thoughts are often encoded under many guises, surely more than one. This book is to an extent inspired by Kurzweil's blueprint for the digital mind. Accordingly, the central problem addressed in the book is what constitutes a model for reality, that is, for the collection of events, in Wittgenstein's *anschauung*.

As time series data emanating from dynamical systems is investigated through the methods of machine learning, a model is said to have been discovered when a sparse set of differential equations on the intrinsic coordinates is identified and the dynamical system can be recovered or fleshed out from the model by means of an autoencoder or a hidden Markov process. Yet, models said to have been "discovered" usually correspond to cases where the answer is known beforehand! What about systems of wanton hierarchical complexity, such as those encountered in biology, biomedical engineering or cosmology?

If the working of the mind serves as any inspiration – and it should since, after all, we are dealing with artificial intelligence (AI) – the learning technologies should first acquire an "intuition," that is, something more primeval than a model, what in this book is termed a *metamodel*. One would say that topology serves as metamodel for geometry, and hence it is only natural that the methods of topological dynamics should be applied to model discovery in dynamical systems studied with artificial intelligence. That is, squarely and simply, the purview of this book, and we shall argue that in most instances a whole hierarchy of metamodels is required to deal with various levels of abstraction in handling the data, a process very much akin to Kurzweil's proposal for the digital mind.

It should be emphasized, with some degree of conceptual violence, that a metamodel is to artificial intelligence what intuition is to the human mind, at least to its biological embodiment.

The leveraging of artificial intelligence for model discovery in dynamical systems is revolutionizing both disciplines, leading to an invigorating cross-fertilization. The proposed book is part of this endeavor, but portends to take the symbiosis to the next level. With the implementation of the topological methods delineated in the book, data science will be able to focus on levels of complexity and multiscale hierarchies hitherto considered off limits. This is so because the dynamic information is encoded at the maximum economy of means, hence greatly simplifying the computations while enabling a decoding of the predictions made at the level of topological dynamics.

As shown in other books in this area, applied mathematicians have been deploying machine learning methods for model discovery in the study of dynamical systems. Yet, in the systems typically showcased – Lorenz strange attractor, reaction-diffusion patterns, and turbulence scenarios in fluid dynamics – the answer is already known, so the "discovery" aspect is in fact missing, as the showcases become in fact test cases for validation. The proposed book truly addresses problems in dynamical systems where the answer is not known but the metamodel can be nevertheless verified as it yields the correct destiny state from within an astronomical number of possibilities.

In dealing with dynamical systems using AI-based approaches, the proposed book addresses the following core question: What constitutes an insightful parsimonious model? The standard answer is: "a sparse system of differential equations on latent coordinates." As this book argues, this is not necessarily the format chosen by AI, given the "dimensionality curse" associated with the complex reality of the systems the book focuses on. The deployment of AI requires a paradigm shift, where dynamic information gets encoded in a topological metamodel. The metamodel is essentially pattern-based, where AI-interpretable topological patterns encode physics laws. These methods are likely to advance model discovery as they enable the reverse engineering of time series stemming from vastly complex realities hitherto inaccessible to the machine learning methods at hand.

The topological dynamics methodology described in the proposed book will render tractable problems in model discovery hitherto considered off limits for AI-based approaches. Thus, ultra-complex hierarchical realities recreating cellular, biomedical, or cosmological contexts will be within reach as the topological methods are incorporated to AI-empowered metamodel discovery. A wide range of applications such as personalized targeted therapy, or cosmological manipulation, not long ago considered part of the fabric of dreams, are in all likelihood materializing far sooner than anyone could fathom. Hopefully, this book has something to do with it.

Ariel Fernández
North Carolina, USA, May 9, 2022

About the Author

Ariel Fernández (born Ariel Fernández Stigliano) is an Argentine-American physical chemist and mathematician. He holds a PhD in Chemical Physics from Yale University and held the Karl F. Hasselmann Endowed Chair Professorship in Bioengineering at Rice University until his retirement. Ariel Fernandez also served as Adjunct Professor of Computer Science at the University of Chicago. To date, he has published over 400 scientific papers in professional journals including *Nature, PNAS, Nature Biotechnology* and *Physical Review Letters.* Fernández has also authored five books on biophysics and molecular medicine and holds several patents on technological innovation. Fernández is a member of CONICET, the National Research Council of Argentina, and heads the Daruma Institute for Applied Intelligence, the research arm of AF Innovation, a Consultancy based in Argentina and the United States.

I

Fundamentals

## CHAPTER 1

Artificial Intelligence and Dynamical Systems

Suppose someone were to say "Imagine this butterfly exactly as it is, but ugly instead of beautiful".

– Ludwig Wittgenstein

1.1 ARTIFICIAL INTELLIGENCE FOR MODEL DISCOVERY

The leveraging of artificial intelligence (AI) for model discovery in dynamical systems is revolutionizing both disciplines, leading to a mutually invigorating cross-fertilization. With the leveraging of AI, dynamical systems have found a fertile ground for development, and, reciprocally, the mathematical theory of dynamical systems is significantly expanding the technological base of AI. As part of this endeavor, this book portends to take the symbiotic relation to the next level in dealing with highly complex realities: metamodel discovery. With the implementation of topological methods, AI-empowered metamodel discovery is able to focus on levels of system complexity and multiscale hierarchies considered off limits in current machine learning (ML) technologies. This is so because the information on time series is encoded at the maximum level of coarse-graining; hence, it greatly simplifies the computations while enabling a decoding of the information generated at the level of a topological description.

Applied mathematicians have been deploying machine learning methods for model discovery in the study of dynamical systems for almost a decade. Yet, in the systems they typically showcase – Lorenz strange attractor, reaction-diffusion patterns, and turbulence scenarios in fluid dynamics – the answer is already known, so the "discovery" aspect is in fact missing: The situations discussed become in fact test cases for further validation. This book addresses problems in dynamical systems, where the answer is not known but the metamodel can be verified as it yields the correct destiny state from within an astronomical number of possibilities. The approach is suited for a vast range of applications, from biological to biomedical to cosmological.

In dealing with dynamical systems using AI-based approaches, we address the following core question: What constitutes an insightful parsimonious model? The standard answer is: "a sparse system of differential equations on latent coordinates." As argued in this and the next chapter, this is not necessarily the format chosen by AI, given the "dimensionality curse" associated with the ultracomplex realities we chose to work on. The deployment of AI requires a paradigm shift, where dynamic information gets encoded in what would be termed a "topological metamodel." The metamodel is essentially pattern-based, where AI-interpretable topological patterns encode physics laws. These methods are likely to advance model discovery as they enable the reverse engineering of time series stemming from vastly complex realities hitherto inaccessible to other machine learning methods.

The topological dynamics methodology described in the book will render tractable problems in model discovery hitherto considered off limits for AI-based approaches. Thus, ultra-complex hierarchical realities recreating cellular, biomedical, or cosmological contexts will be within reach as the topological methods are incorporated to AI-empowered metamodel discovery. These advances represent substantial contributions to dynamical systems research and have implications for a vast array of applications. Such applications range from targeted therapy and molecular medicine, where the assessment of the *in vivo* reality is crucial to discover therapeutic agents capable of efficacious molecular intervention (Chapters 3 and 4), to cosmology, where the physical underpinnings of quantum gravity may stem as emergent properties in a statistical thermodynamic treatment of machine learning systems (Chapter 5).

1.2 PRIMER ON DEEP LEARNING

Artificial intelligence refers to machines capable of exhibiting behavioral traits that humans regard as indicators of intelligence, such as learning and problem-solving [1]. Within this protean and fuzzily delineated subject, machine learning refers to the ability to learn without being explicitly instructed to do so, while deep learning (DL) refers to an automated extraction of features, patterns, and ultimately models from arrayed data that is sequentially represented within an abstraction hierarchy organized as a multilayered neural network (NN) [2–4]. DL will be the aspect of AI that this book mostly focuses on as we seek to unravel mathematical models enshrined in the patterns that underlie vast arrays of dynamic data.

DL has been shown to be highly efficacious at identifying features that are in principle discoverable from the data [2, 5]. As in face recognition, features are hierarchically organized, so that large-scale patterns (eyes, noses, face shapes) emerge after several layers of abstraction from simpler or more rudimentary patterns (lines, curves, shades). The beauty and power of DL resides in the fact that the feature extraction process may be carried out in an unsupervised manner: the features emerge from the training of the system without human input or bias, and enable the network to make accurate inferences. In this era of big data, arising primarily in biology, biomedicince, and cosmology, we may state that time is ripe to plunge into DL and master the field. There are several compelling reasons for taking the DL approach:

- Fields like biology, biomedicine, and cosmology are generating vast amounts of data and time series that can be easily and inexpensively stored and require interpretation and ultimately conceptual unification within overarching models.

- We have the right hardware, that is, graphics processing units (GPUs), that are massively parallelizable.

- We have adequate software such as TensorFlow (TF) that enables suitable modular coding if the data can be pixelized or voxelized into a tensorial array, be it a vector, a matrix, or a tensor proper [3, 4].

At the most basic level, the building block of an NN is the neuron, referred to as *perceptron* [5]. The perceptron enables forward propagation of information encoded in an array of inputs $x_1, x_2, \ldots, x_n$ weighted by parameters $w_1, w_2, \ldots, w_n$ to generate an output of the form $y = f\left(\sum_{i=1}^{n} x_i w_i + w_0\right)$, where w_0 may be regarded as a bias term and f is a nonlinear activation function, often a sigmoid or sigmoid shaped, as shown in Figure 1.1. The bias term enables to shift the activation function. In vector representation, we often write: $y = f(z)$, $z = x.w + w_0$, where x,w are, respectively, input and

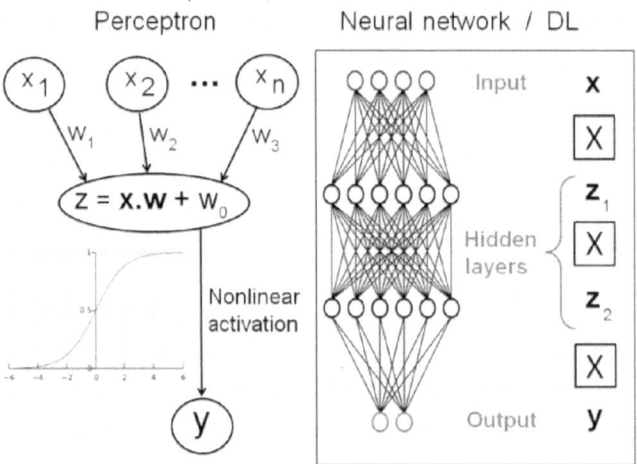

FIGURE 1.1 Scheme of the perceptron or neuron activation by linear transformation of input stimulus (x) followed by nonlinear signal transmission as output y. The panels on the right show the organization of NNs with one and multiple hidden layers. The "boxed X" indicates full (dense) node connectivity between consecutive layers.

weight vector. One can subsequently build a fully connected (dense) layer of perceptrons indexed by j, whereby $z_j = x.w_j + w_{0,j}$. The dense layer can be readily implemented in TF code, by simply specifying number of outputs/perceptrons [3, 4]. We usually refer to the vector of linearly transformed inputs z as "hidden layer," as it does not explicitly describe observables [5].

Hidden layers may be stacked as in DL architectures. Thus, for the j-th perceptron in the k-th hidden layer, we get: $z_j^{(k)} = \sum_{i=1}^{n_{k-1}} y_i^{(k-1)} w_{ij}^{(k)} + w_{oj}^{(k)} = \sum_{i=1}^{n_{k-1}} f\left(z_i^{(k-1)}\right) w_{ij}^{(k)} + w_{oj}^{(k)}$. The sequential composition of the network by stacking hidden layers has a standardized script in TF that generates the propagation of information as specified by the equation above [3, 4]. In an NN with K hidden layers, we may regard the output $y_j^{(K)} = f\left(\sum_{i=1}^{n_{K-1}} y_i^{(K-1)} w_{ij}^{(K)} + w_{oj}^{(K)}\right)$ as an "inference" made by the DL system.

The accuracy of the DL inference represents the level of optimization of network performance and may be assessed vis-à-vis a training set of input/output paired datapoints. This assessment is often referred to as the *loss* of the network [2, 5]. The loss is parametrically dependent on the full weight tensor $W = [w_{ik-1jk}^{(k)}]_{k=1,\ldots,K}$, which may contain a huge number of weights, in the thousands, if not millions, depending on the size of the network. Thus, for DL NN with K hidden layers the loss function or empirical risk $J(W)$ becomes $J(W) = |\mathfrak{I}|^{-1} \sum_{\xi \in \mathfrak{I}} \mathcal{L}\left(y_\xi^{(K)}(W), y_\xi\right)$, where $\mathfrak{I}$ is the training set with number of elements $|\mathfrak{I}|$, $y_\xi^{(K)}(W)$ is the predicted output vector for input vector x_ξ, $\xi \in \mathfrak{I}$, y_ξ is the actual output vector, and $\mathcal{L}\left(y_\xi^{(K)}(W), y_\xi\right)$ measures the discrepancy between actual and predicted output. In regression problems, where the output is a numerical vector, it is often convenient to adopt $\mathcal{L}\left(y_\xi^{(K)}, y_\xi\right) = \|y_\xi^{(K)}(W) - y_\xi\|^2$. In such cases, the optimization of the NN through training becomes a problem of least squares. To optimize the NN is tantamount to minimize the loss $J(W)$, which requires a careful fine-tuning of the size of the training set vis-à-vis the size of the weight parametrization. In principle, the optimal network is defined as: $W = W^* = Arg \min J(W)$.

An insufficient number of training input/output pairs relative to the size of the weight tensor would give rise to *overfitting*, requiring special techniques, generically known as *regularization*, in order to trim the network, that is, randomly remove connections, without compromising predicting efficacy [2, 5].

Optimizing the network is no trivial task, as it involves the laborious and costly computation of the minus gradient $-\dfrac{\partial J(W)}{\partial W}$, which locally indicates the direction of steepest descent in the multidimensional surface $J = J(W)$. An iterative gradient descent computation generating a fine-tuned weight updating $W \rightarrow W - h\dfrac{\partial J(W)}{\partial W}$ should eventually lead to convergence to a local minimum of $J = J(W)$ when adopting a suitable learning step h. This parameter should be tuned to effectively escape local minima while avoiding overshooting in trying to reach the global mimimum. Most gradient descent algorithms use an adaptive learning step during training, in accord with the constraints indicated. In practice, the gradient problem is approached by what is called the *stochastic gradient descent* method, whereby not all datapoints (input/output pairs) in the training set are used in each minimization step but the gradient is approximated by an average over randomly chosen batches of datapoints in a trade-off between accuracy and computational efficiency. Obviously, batch size and learning rate are correlated, so the more accurate the gradient estimation, the larger the learning step may be (a token of computational confidence). To achieve significant speed, the stochastic gradient descent computation may be massively parallelized by splitting up batches into multiple GPUs.

1.3 NEURAL NETWORKS AS MODELS FOR DYNAMICAL SYSTEMS

Throughout this book, we shall be mostly (but not exclusively) concerned with data organized as a time series arrayed as $\{x(t_0), x(t_0 + \tau), x(t_0 + 2\tau), \ldots, x(t_0 + L\tau)\}$, where $x(t)$ is the vector of observables at time t and τ is the interval that determines the time coarse-graining inherent to the sequential detection registered in the vector x. The time series enables a training of the NN such that the output vector $y = y(x(t), W)$ should approximate $x(t + \tau)$ when the input is $x(t)$, and this correspondence is carried over all t in the training time series. Thus, for a given network architecture and time series, the optimal network is the one that realizes the minimum of the loss function:

$$J(W) = \sum_{n=1}^{L} \|x(t_0 + n\tau) - y(x(t_0 + (n-1)\tau), f, W)\|^2 \qquad (1.1)$$

Obviously, for a fixed activation function, a model for the time series may be given simply by $Arg \min J(W)$, but such a model would lack universality as it would be extremely parametrized, most likely overparametrized, and would not prove insightful in the sense that it is not parsimonious. Discovering a suitable model for a time series is the overarching goal of this book. The discussion prompts us to enquire what truly constitutes a model. This question will be addressed in Chapter 2.

The most common time series that humanity has collected since time immemorial stems from astronomical observation. For about two millennia until the time of Copernicus, and Newton later on, humanity had been striving to find a suitable model that would fit and explain the data, that is, the recorded sequential positions of a set of celestial bodies. In today's more general context, biology, biomedicine, and cosmology are generating dynamical data at a staggering rate, while models that fit and explain the data are sorely lacking or hopelessly irrelevant and hence inconsequential. It is expected that the advent of AI will turn the tables around and dramatically impact model discovery. In essence, we seek for what is known as autoencoder, an intermediate output with a drastic dimensionality reduction and a simplified discerning physical picture that should therefore prove insightful to make sense of the patterns enshrined in the time series, thus enabling meaningful output inferences. These autoencoders and the models they give rise to will be studied in detail in Chapter 2 and illustrated in the subsequent chapters.

1.4 DEEP LEARNING WITH BIOMEDICAL APPLICATIONS

Vast amounts of biomedical and biostructural data have become the hallmark of the post-genomic era, and the sheer size of chemical combination possibilities makes it forbiddingly difficult to parse chemical space in search for suitable leads for targeted therapy [6]. In this scenario, it is only natural that pharma researchers have turned to DL for guidance in the drug discovery and development process [7–10]. Thus, areas like cheminformatics or pharmacoinformatics have benefited immensely from the advent of DL systems trained by pairing chemical compounds with molecular attributes that are relevant to targeted therapy. These computational and informational techniques enable the evaluation of chemical compounds for specific properties, including target affinity, affinity screening profiles, structural features of drug-target docking, and absorption, distribution, metabolism, excretion, and toxicity (ADMET) profiling [8].

A major challenge in implementing DL models for pharmacoinformatics in accord with the generic scheme outlined previously stems from the need to represent the chemical structure as a tensor array (vector, matrix, or tensor proper) of pixelated or voxelated inputs that may be subsequently interrogated geometrically across the hidden layers of NN in search for features that are indicative of the molecular properties indicated. There are a number of representations of chemical space amenable to TF encoding. The most obvious one is to order the atoms in the compound on a 1D-array (following, e.g., the IUPAC numerical labeling convention) and represent the chemical structure of the molecule as a covalent bond matrix pairing atoms in row and column in accord with their covalent linkages, including single, multiple, and resonant (aromatic) bonds. The matrix is subsequently transformed into a topological descriptor that describes the invariants arising from different atom ordering. The input describing chemical structure is paired within a training set against the molecular attributes of therapeutic relevance that the network is meant to infer. Then, compounds that need to be evaluated/profiled are inputted as the array of pixels/voxels, transformed into a topological representation, and feature extraction is achieved through the sequential activation of hidden layers at increasing levels of abstraction, eventually leading to the profile inference which is subsumed in the output layer.

1.5 CONVOLUTIONAL NEURAL NETWORKS IN DRUG DESIGN

Feature extraction through the NN often requires particular architectures known as convolutional NNs (CNNs) [2–5]. The idea is to pixelize or voxelize the input data in a matrix or 3D tensor array and then scan (convolve) the array with a filter associated with a specific pattern to generate feature map. For the sake of the argument, let us consider a 2D-array input M. The filter $F = (w_{ij})$ is an mxm matrix actually representing a convolutional kernel, so that a neuron in the F-associated hidden layer M_F only senses the pixels in an mxm patch (receptive field), and the layer becomes the feature map

$$M_F = M * F = \left(\sum_{j=1}^{m} \sum_{i=1}^{m} w_{ij} x_{i+ap\,j+bp} + w_0 \right)_{ab}, \qquad (1.2)$$

where p, usually set at $p = 2$, is the stride adopted as the filter slides along the input matrix array, with dummy integers a, b indicating the patch location. Essentially, the CNN is an NN where the set of weights in each mxm patch of inputs are always the same as the filter slides along the input array to generate the hidden layer that constitutes the feature map. The convolution operation becomes the entry-by-entry (Frobenius) matrix inner product as the filter slides along the input matrix with a given stride. Successive filters may be applied reducing progressively the size of the feature maps as higher and higher levels of abstraction are achieved in the representation of the data (Figure 1.2). Thus, the parametrization required for feature extraction in CNNs is relatively small, as all the neurons in a hidden layer partake the same connectivity parameters that define the filter that generated that layer.

The "beauty and magic" of CNN lies in the fact that the filters are not specified a priori, that is, their weights are not fixed through human intervention. The filter parametrization is automatically determined by the training/optimization process without introducing any assumption, other than the size of the receptive fields for each convolution operation and the overall number of filters to be applied. Thus, feature extraction in a CNN is carried out in an unsupervised manner and only requires that we script (in TF coding) the number of hidden layers or feature maps and the size of the filters to be applied to generate each hidden layer. The features themselves emerge as the network is trained.

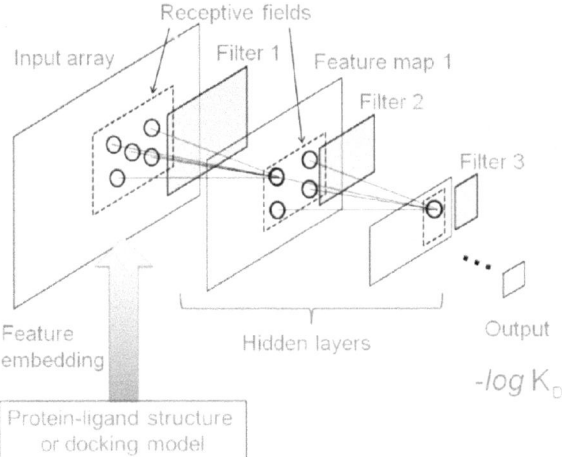

FIGURE 1.2   Scheme of a CNN used for inference of drug-target affinity through sequential application of filters that become optimized through minimization of the network loss.

Thus, CNNs often constitute AI-empowered platforms for drug discovery, where the structure of a protein-ligand interface that serves as precursor for a predicted drug-target interaction is pixelated in a feature embedding process as a 3D spatial array of protein-drug atom pairs deemed to be interacting across the interface [11]. The inference of target affinity for a given drug is assessed through a sequence of feature extractions using convolution filters until the output feature map becomes a number directly associated with the affinity $pK_D = -logK_D$, where K_D is the dissociation constant for the drug-target complex inputted via feature embedding (Figure 1.2). The training of the network is carried out by minimizing the network loss or empirical risk over a set of drug-target complexes whose structure and affinity are both known (preferably, experimentally determined). The proteins in the complexes of the training set are typically homologs of the one whose affinity for a specific drug we seek to infer, so that the features that enable the affinity inference may emerge from structural alignment.

As shown in Chapter 4, in more complex applications of CNNs, where the output is not simply a numerical parameter but rather a numerical array, it is often convenient to adopt a variation of the CNN architecture in which the patches that yield the pixels with the highest weights in the feature map are reconstituted into features perfectly stenciled by the filter, while other patches that do not yield discernible features are left invariant

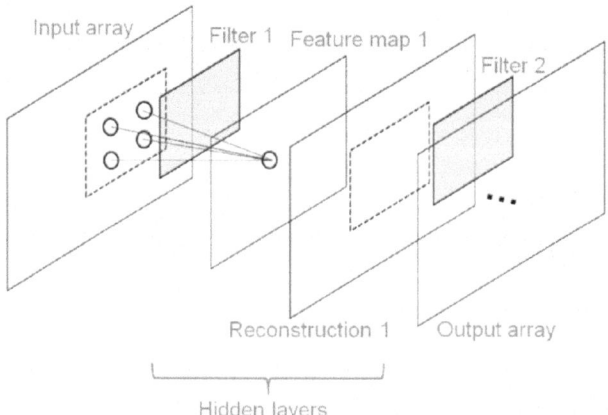

FIGURE 1.3 Scheme of a CNN with a particular type of architecture in which the patches that yield the pixels with the highest weights in the feature map are recreated as filter-stenciled "perfect" features. Thus, a feature-enriched reconstitution of each layer is carried out prior to the application of the next filter.

in nonoverlapping regions. Thus, with the application of each filter, a feature-enriched reconstitution of the input layer is carried out prior to the application of the next filter (Figure 1.3).

1.6 DEEP LEARNING FOR DYNAMIC TARGETS IN BIOMEDICINE

As previously highlighted [7–11], forays in ADMET and pharmacokinetics and pharmacodynamics (PK/PD) profiling, structural/affinity predictions on drug-target associations, virtual screening and other cheminformatics or pharmacoinformatics applications are all successfully exploiting AI, and in particular DL, to develop useful predictors. In these studies, drug and target protein are represented as rigid entities not influencing each other's conformation, so the dynamics of the induced folding or structural adaptation of the target protein to create an optimal drug-target interface [6] is absent altogether from the data analysis.

Therapeutic agents target dysfunctional or deregulated proteins in order to disable them, but the target itself is often a moving target, as protein regions prone to structural adaptation are typically the ones that bind to the drug [12]. Furthermore, the DL-compatible pixelated representation of the drug-target interface typically assumes pairwise interactions [11], but we now know that protein associations involve many-body energy terms: water exclusion from around preformed hydrogen bonds in one of the binding partners upon drug-target association plays a key role in driving the interaction [6]. Thus, AI-driven forays to investigate drug-target associations require a new discipline named epistructural biology [6]. As currently defined and understood, structural biology falls short of the proper conceptual/representational background to enable AI-driven forays into the design of drug-target interfaces [12]. Thus, a detailed representation of the dynamic drug-target interface incorporating many-body terms arising from water exclusion needs to be streamlined into the tensor flow of DL systems to bring drug design to the next level, that is, mastering the engineering of drug-target interfaces and the dynamic problems arising thereof.

Last but not least, drugs target dysfunctional proteins in an *in vivo* setting, not an *in vitro* setting, while proteins are selected to fold efficiently *in vivo*, not *in vitro*, and the drug-induced folding upon binding occurs *in vivo*, not *in vitro*. Thus, the *in vivo* context needs to be factored into the dynamic analysis used to engineer the drug/target interface. At this juncture, AI and only AI can steer intuition and provide guidance to design therapeutic drugs and engineer drug-target interfaces *in vivo* by mimicking or simulating the cellular setting where such interfaces occur.

In Chapter 4, we deploy methods based on AI to tackle dynamic problems in drug design, including the incorporation, mimicking, and simulation of *in vivo* settings. In so doing, the arsenal of epistructural biology is introduced to provide the proper representational and conceptual framework to enable inference through the tensor flow of CNNs.

1.7 DEEP LEARNING HAS SOLVED ONE OF THE PROTEIN FOLDING PROBLEMS

On November 30, 2020, DeepMind, a Google company, announced that their AI system AlphaFold2 had solved the "protein folding problem" [13, 14]. That is, by adopting a deep learning CNN trained on structural data, and correlating it with evolutionary input, AlphaFold2 is capable of predicting the 3D structure of a protein chain from its amino acid sequence. This constitutes a towering achievement because the protein folding problem is the biggest challenge in biology or biophysics, has remained an open problem for over 50 years, and has been deemed by most scientists as almost intractable. Perhaps the most astonishing and paradoxical aspect of this achievement is that the expertise profile of the team that developed AlphaFold does not resemble even remotely that of the researchers that have worked (almost fruitlessly) on the problem for the last 50 years.

Thus, molecular biophysics is the most underrepresented field in the team's expertise, and yet the problem they have solved is one of the paramount problems in molecular biophysics.

To solve the paradox, we need a better grasp on what DeepMind has truly done. They have not tried to predict protein folding pathways that converge to the native fold as the protein chain explores conformation space governed by physical principles. That effort, which has distracted biophysicists for decades, is in all likelihood completely misplaced and misleading. Natural proteins are selected to fold *in vivo*, not *in vitro*, and current computational technologies are completely at a loss in terms of modeling the complexities of an *in vivo* environment. This situation is likely to change with the advent of AI, as clearly demonstrated in Chapter 3.

By contrast, DeepMind has trained the AlphaFold CNN with plenty of structural information on native structures from protein data bank (PDB) entries, hence subsuming in those patterns all the physics and complexity of the *in vivo* environments. In other words, the relevant information is enshrined in the training set without requiring anyone developing the CNN to know about it. Then, through sequence alignment, the test sequence is evolutionarily related with sequences associated with structures in the training set, and the result of the alignment is inputted into the CNN along with the test sequence itself. The beauty and magic of the AI system is that it is able to infer the structure without explicitly learning the physics that governed the formation of the structure. In fact, it is most likely that the feature extraction in the CNN would reveal a completely novel, and hopefully perfectly equivalent, set of physical principles tailored to the different levels of representational abstraction in the hidden layers. The features extracted in the hidden layers probably enshrine a novel and fascinating set of physical principles optimally tuned to the biology they govern. They are just waiting to be discovered.

1.8 AI-EMPOWERED METAMODEL DISCOVERY FOR HIERARCHICAL DYNAMICAL SYSTEMS: ADIABATIC REGIMES, LATENT MANIFOLDS, QUOTIENT SPACES

It is widely expected that AI and DL in particular will become a major player in model discovery for data-driven research [15]. Within this vast array of possibilities, the focus of this book is model discovery for dynamical systems that underlie big time series data from vast areas as distant as biomedicine and cosmology [16–18]. The biggest hurdle we stumble upon is that the level of complexity of the data generated in fields like biology and cosmology is so extreme that it forces us to challenge the notion of model itself. As shown in the subsequent chapters, particularly in Chapter 2, the model itself can seldom be cast as a system of differential equations. Thus, much of the ensuing discussion deals with the question of what constitutes a meaningful model with predictive value when dealing with the ultracomplex realities represented in the contexts of biology or cosmology.

Encoding has proven to be a necessary category in AI [15–18]. The type of AI we are mostly concerned with involves deep learning, which requires an encoding of the raw information that needs to be acquired and processed further to make meaningful inferences. Just like with human intelligence, the encoding problem stems from the core question: What is essential and what it superfluous? The encoding problem becomes solvable when the system under scrutiny is hierarchical and the hierarchical structure is fairly obvious or at least discoverable. In the cases treated in this book, the structure of the data is always hierarchical, implying that the learning process admits a reductive approach represented by the encoding. When the network architecture is such that this process is automatically generated, we name the NN *autoencoder*.

We shall deal mostly with time-dependent data representing physical or biophysical systems, where detailed fast motions may be systematically averaged out, so the relevant information may be stored in a coarse-grained representation. Rigorous mathematical constructs will be introduced to implement the hierarchical encoding materialized by the autoencoder. Some examples of hirerarchical encoding of physical or biophysical processes that needs to be taken into account when designing the autoencoder architecture are as follows:

1. The adiabatic approximation, where fast-relaxing or fast-evolving enslaved degrees of freedom are averaged out, or thermalized, or equilibrated when incorporated into a model for the time evolution of a dynamical system [19]. In molecular physics, examples of such degrees of freedom are vibrational hard modes that evolve on timescales of the order of picoseconds to nanoseconds, while soft modes evolve on longer timescales ranging from the submicrosecond to seconds.

2. In atomic physics, the Born–Oppenheimer approximation represents regions of the potential energy surface (PES), where an adiabatic regime holds so the motion of electrons is enslaved or entrained by the slower motion of the atomic nuclei [20].

3. The latent manifold in dynamical systems [15], where the system is entrained or enslaved in the long-time limit by the evolution in a manifold of lower dimension than the original space. The latent manifold is often referred to as center manifold [19], especially in the context of dissipative systems, and contains the attractors of the system (Figure 1.4).

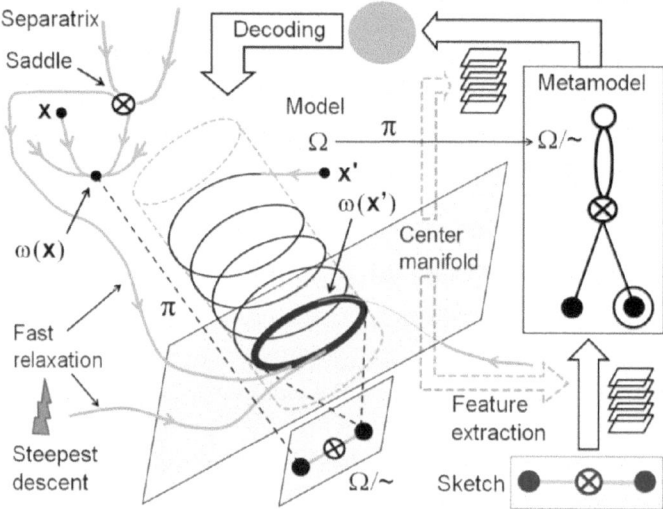

FIGURE 1.4   Schematics of the topological "metamodel" encoding of a dynamical system that contains a center manifold. The center manifold enslaves or entrains the dynamics for timescales associated with the adiabatic elimination of fast-relaxing and thermalized degrees of freedom and therefore constitutes a latent manifold (Ω) within which a model (differential equations on latent coordinates) may be identified by an autoencoder. The dynamic information may be encoded further by a second autoencoder at a higher level of abstraction, where the dynamics are represented more coarsely as "modulo basin" transitions. The modulo-basin dynamics is mapped on a quotient space, $\Omega/\sim$, where two states *x, y* are regarded as equivalent, *x~y*, if they have the same destiny state ($\omega(x) = \omega(y)$). The encoding processes are symbolized by dashed lines, and the enslaving center-manifold dynamics is highlighted by a thick dark circle. Thus, the second autoencoder materializes the projection $\pi : \Omega \rightarrow \Omega/\sim$, and represents the dynamics as a walk in a graph whose vertices are the critical points (minima, saddles of different indices, maxima) and attractors of the system, and the edges represent connections along pathways of steepest descent. This topological metamodel may be subsequently decoded back to flesh out the dynamical system using learning technology.

4. In the biophysical context of functionalized soluble proteins (enzymes), the quantum mechanics for specific chemical processes takes place at the *epistructure* (solvent organization around the protein structure), where water is chemically functionalized [21]. A rigorous treatment of this problem would require that we solve a time-dependent Schrödinger equation. In this context, an AI approach would need to be incorporated to learn to infer the nodal structure of molecular orbitals within a voxelated 3D grid. With current technologies, this AI approach is plausible only under the adiabatic regime given by the Born–Oppenheimer approximation.

5. In clonal population dynamics, cancer phenotypes become selected under therapeutic pressure [22]. This dynamics is complex but hierarchical, and the mathematical procedure to "encode what is essential" in this context is based on the center manifold reduction (Figure 1.4). This reduction, discussed in (3), is actually a projection of the dynamical system onto a lower dimensional system that entrains it. The autoencoder may discover the center manifold by interrogating vast amounts of time-dependent data, as it seeks to meaningfully reduce dimensionality in such hierarchical systems.

6. Chapters 2 and 3 introduce *quotient space*, a fundamental mathematical construct to take advantage of dynamic hierarchy in order to encode information as required to implement DL systems [23]. The quotient space is built upon an underlying dynamics and may be equated with the orbit space, that is, points in the same trajectory are regarded as equivalent, and the quotient space is the set of equivalence classes with a topology inherited from the underlying space where the dynamical system is mapped (Figure 1.4). In rigorous terms, two points-states (x, x') are equivalent $(x \sim x')$ or belong to the same equivalence class $(x' \in \bar{x}, x' = \bar{x})$ if and only if they share the same destiny state $(\omega(x) = \omega(x'))$ vis-à-vis the trajectory to which they both belong. Thus, we simplify the space by lumping microstates into basins of attraction (of destiny states) in the potential energy surface. The modulo-basin dynamics constitutes a coarse model named *metamodel*, which is far easier to encode as the system learns data generated by atomistic molecular dynamics (MD) simulations. In this way, AI learns to propagate dynamics in quotient space, discovering a metamodel to cover physically realistic timescales usually inaccessible to detailed atomistic computations.

While the center manifold encoding is suited for dissipative dynamics, where the attractor may be nontrivial (cf. Figure 1.4), the quotient space simplification is better suited for Hamiltonian systems. Both levels of encoding converge as free-energy dissipation tends to zero. In fact, as the AI system encodes time-dependent raw data, it implicitly composes the two levels of encoding, with the center-manifold reduction averaging out of fast modes, followed by projection onto quotient space (Figure 1.4).

In all cases dealt with in this book, the hierarchy of the data that makes it amenable to encoding either is AI-discoverable (Chapter 3) or may be unraveled through a rigorous mathematical construct that needs to be incorporated to the learning code (Chapter 2). In the future, AI systems will likely be able to unravel such hierarchies without human intervention and engineer the multilayered NN accordingly. Conceivable, in the future AI may also strive on systems that lack hierarchy, while it is doubtful that the human mind can ever handle such cases.

The modulo-basin hierarchical representation of the dynamics in quotient space enables the construction of metamodels, that is, coarse-grained models that represent transitions between equivalence classes of states of the system, essentially reproducing the topological dynamics within a graph representation (Figure 1.5). As an illustration, let us consider the "keto" $\rightleftarrows$ "enol" interconversion of acetone (Figure 1.5). This is a chemical reaction involving an intramolecular proton migration that can be modeled by the potential energy surface representing the ground-level electronic energy sheet under the Born–Oppenheimer approximation. From quantum mechanics calculation, we know that this PES has three minima, representing the less stable "enol" form (1), the more stable "keto" form (2), and a state where the jumping proton is completely dissociated from the rest of the molecule (3). The minima are separated by four saddle points at the top of the path of steepest descent joining pairs of minima on both sides of the saddle along the direction of negative curvature. This direction is actually the eigendirection associated with the eigenvalue of the Hessian matrix (Jacobian of the gradient vector field) with negative real part computed at the saddle point. In turn, the lines of steepest descent joining the four saddles with the three maxima become the separatrices of the basins of attraction of the minima. Thus, all lines of steepest descent may be represented as a graph (metamodel II, Figure 1.5) joining critical points that are in turn organized in tiers, where the lowest tier corresponds to points where all eigendirections have positive curvature

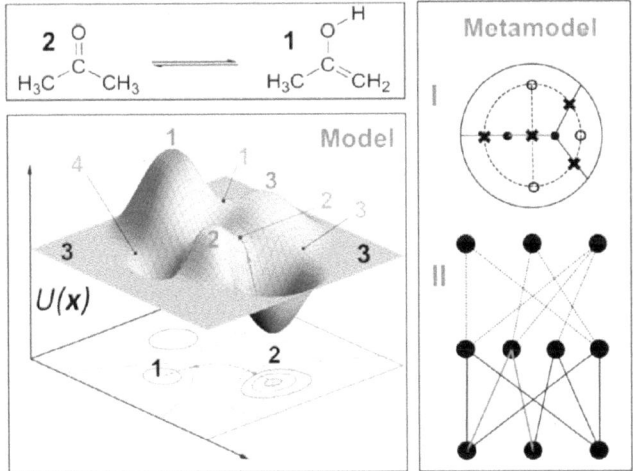

FIGURE 1.5 Three levels of abstraction in the modeling of the chemical dynamics for the isomerization "keto" $\Longleftrightarrow$ "enol" of acetone, representing the intramolecular migration of a proton. The potential energy surface (PES) represents the model and the topological relations between critical points, together with the "modulo basin" transitions along the paths of steepest descent joining critical points represent the metamodel. On metamodel I, the modulo-basin topology is mapped onto the latent manifold, while on metamodel II the modulo-basin topology is abstracted further, and displayed as a graph.

(minima) and the next tier includes all critical points with only one eigen-direction with negative curvature (saddles of index 1); the next tier is associated with two eigendirections of negative curvature (maxima in the case of Figure 1.5), and so on. The edges on the graph represent paths of steepest descent joining critical points in adjacent tiers. In this way, the topological dynamics of the PES may be encoded as a metamodel represented in graph form, while the reversible chemical reaction pathway becomes a walk in the said graph.

As argued in the subsequent chapters, metamodels enable the discovery of hierarchical dynamical systems underlying processes that unravel in realities of high multiscale complexity. The implementation of a metamodel factorization of the dynamics requires the concerted participation of several components operating in a coordinated manner within an AI platform. First, we need to introduce the so-called autoencoders that constitute the deep NN systems that encode the dynamics on the "latent manifold" Ω. This manifold entails a significant dimensionality reduction relative to state space W and is spanned by the internal coordinates that label the orbits of the symmetry group $\mathcal{G}$ inherent to the system and acting on W: $\Omega \approx W/\mathcal{G}$. Traditionally, it is expected that the first autoencoder, hereby denoted AE1, that generates the latent manifold is jointly optimized to generate also the most sparse or minimal set of differential equations on the manifold that can be decoded back onto the dynamical system defined on W. This is what is usually meant by "model discovery." In practice, the level of multiscale complexity of the processes dealt with in this book does not make models amenable to discovery at the geometric level. As previously argued, in such cases another level of abstraction $\pi : \Omega \rightarrow \Omega/\sim$ needs to be introduced so that the coarse graining of time within $\Omega/\sim$ reflects equilibration within the basins of attraction in the latent dynamics. This hierarchical escalation in the level of abstraction requires a second autoencoder, AE2, capable of encoding the topological features of the latent vector field.

At this stage, a different sort of NN architecture is required to propagate the metadynamics on $\Omega/\sim$. To properly delineate the architecture of the NN required for metadynamic propagation, we limit the discussion to the case where the dynamics is generated by a smooth (i.e., C^1) potential energy function $U : W \rightarrow \mathbb{R}$ invariant upon the isometries (distance-preserving transformations) of W. Furthermore, $\mathcal{G}$ is an Euclidean group, so that Ω is compact, and hence $\Omega/\sim$ becomes a discrete set of basins, and a basin assignment represents a coarse state of the system. Then, the metadynamics may be encoded as "evolving text" representing a time series of basin transitions. Chapter 3 is primarily devoted to construct and exploit this particular type of metamodel. The textual processing requires the implementation of a particular type of DL architecture known as *transformer*, while the transformer-based propagation of the metadyamics constitutes a Markovian process (Figure 1.6). Through AE2, this metadynamics is decoded as latent dynamics on Ω and validated by contrasting the latent dynamics against the hidden Markov process upheld under the adiabatic conditions described.

Thus, to implement a metamodel within an AI platform, it is essential that the two autoencoders and the transformer are optimized to work concertedly with complete compatibility (Figure 1.6). This means that an input state yields the same destiny state regardless of the pathway chosen, provided the pathways have identical endpoint spaces in the commutative diagram presented in Figure 1.6.

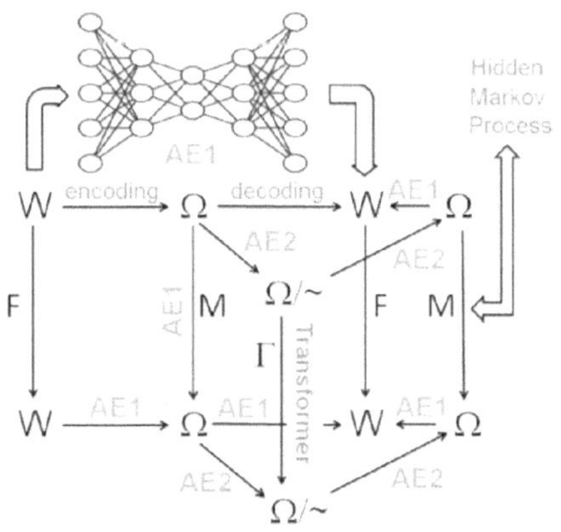

FIGURE 1.6 Commutative diagram representing two coupled autoencoders operating sequentially in tandem and representing two levels of abstraction of a dynamical system. The first autoencoder, labeled AE1, projects the dynamics onto the latent manifold, while autoencoder AE2 projects the latent dynamics onto a discretized "modulo basin" version where a coarse state of the system is represented in textual form and the "modulo basin" dynamics may be learned and propagated within a special type of autoencoding architecture known as "transformer."

1.9 METADYNAMICS FOR METAMODELS: MAPPING OUT THE QUOTIENT MANIFOLD WITH A DEDICATED AUTOENCODER

Here we are concerned with dynamical systems governed by a potential $U = U(x)$ factorized through the latent manifold Ω. The time series may be generated operationally through molecular dynamics or a Monte Carlo process for a limited period of time [23]. The key issue addressed in this section is how to construct the quotient space $\Omega/\sim$ that requires mapping out the basins of attraction in the potential energy surface that underlies the latent dynamics. It has been amply documented that trajectories in latent coordinates generated with state-of-the-art computation technologies cannot effectively surmount sampling bottlenecks [23, 24]. This masking of ergodicity renders them ineffective to generate the quotient space. However, a gamut of computational techniques is now available to enhance the sampling rate and circumvent kinetic traps. Among them, metadynamics stands out for its simplicity and efficacy at "nudging" the trajectory away from previously sampled states by introducing a potential energy bias that depends on the sampling history [24]. Inspired by this procedure, we introduce our own "pseudo-ergodic metadynamics" which is tailored to our need to construct the quotient space that underlies the metamodel. This undertaking entails mapping out the basins of attraction in the potential energy surface or, more generally, the cross sections of the PES topography when we deal with higher dimensional cases.

Let us assume that Ω is compact, as it is the case in most applications dealt with (Chapter 3). Then, for an arbitrarily small $\delta > 0$, there exists a finite δ-cover of the manifold $C(\delta) = \{A_j\}_{j=1,...,N(\delta)}$ consisting of open sets subject to the condition: $\forall j : 0 < diameter(A_j) < \delta$. Based on the finite cover, we may construct a partition $\mathcal{P}(\delta, \Omega)$ of Ω, namely $\mathcal{P}(\delta, \Omega) = \{Q_j\}_{j=1,...,N(\delta)}$, satisfying the defining conditions: (a) $\Omega = \mathrm{U}_{j=1}^{N(\delta)} Q_j$ and (b) $\forall i \neq j$: $Q_i \cap Q_j = \varnothing$, $i, j = 1, ..., N(\delta)$. To construct the partition, we define

$$Q_1 = \overline{A_1}, Q_2 = \overline{A_2} \setminus \left(\overline{A_1} \cap \overline{A_2}\right), Q_3 = \overline{A_3} \setminus \left[\left(\overline{A_1} \cap \overline{A_3}\right) \cup \left(\overline{A_2} \cap \overline{A_3}\right)\right], ..., \quad (1.3)$$

where the bar on top of a set denotes its topological closure, that is, the incorporation of all limit points.

Since, by definition, an ergodic trajectory passes arbitrarily close to each point in Ω at least once, we may safely assume that it visits at least once each component of $\mathcal{P}(\delta, \Omega)$. For a canonical ensemble, we may define the entropy $S(Q), Q \in \mathcal{P}(\delta, \Omega)$ up to an immaterial constant as the thermodynamic limit:

$$S(Q) = T \lim_{t \to \infty} lnM((Q, t)), \quad (1.4)$$

where $M((Q, t))$ is the multiplicity of Q vis-à-vis a generic trajectory $\{x(t)\}$. The multiplicity is defined by:

$$M((Q, t)) = \int_{t'=0}^{t'=t} \chi_Q(x(t')) dt', \text{where } \chi_Q(x) = 1 \, if \, x \in Q \text{ and} \quad (1.5)$$
$$\chi_Q(x) = 0 \text{ otherwise}$$

To accelerate the sampling of the latent manifold, we introduce a history-dependent potential energy bias that "discourages" the trajectory from visiting a coarse-grained state $(Q \in \mathcal{P}(\delta, \Omega))$ that has been visited before. This implies that we are introducing a potential energy bias $V(Q, t)$ at the rate

$$\frac{d}{dt}V(Q,t) = w_0 e^{\beta\left[E^{\#} - V(Q,t)\right]} \chi_Q(x(t)) = w e^{-\beta V(Q,t)} \chi_Q(x(t)), \quad (1.6)$$

where $\beta = (k_B T)^{-1}$ with k_B = Boltzmann constant, $E^{\#}$ is the overall activation energy barrier of the dynamics governed by $U(x)$, w_0 is the pre-exponential kinetic factor, and $w = w_0 e^{\beta[E^{\#}]}$ is the initial and maximum rate of nudge accretion. It should be emphasized that the slower the rate of interbasin transition within the PES, the faster the accretion of potential energy nudges and vice versa. The rate of nudge incorporation decreases in inverse relation to the time-dependent maximum rate $\sim e^{-\beta[E^{\#} - V(Q, t)]}$ of interbasin transition in the PES. Thus, the potential bias contributes more when it is more needed to enhance sampling of the latent manifold: *the steeper the potential well, the bigger the buildup of nudges it gets* (Figure 1.7). We name this procedure pseudo-ergodic metadynamics. Integration of the equation yields the following history-dependent potential bias:

$$V(Q,t) = \beta^{-1} \ln\left[1 + w\beta M(Q,t)\right] \quad (1.7)$$

Thus, the potential $\int_Q U(x)dx + V(Q, t)$ governs the coarse-grained trajectory projected onto the partition $\mathcal{P}(\delta, \Omega)$, which approximates a pseudo-ergodic trajectory on Ω in the limit $\delta \to 0$. The beauty of this sampling-enhancing procedure is that the original PES topography, and thereby the quotient space $\Omega/\sim$, may be directly recovered from the cumulative bias potential in the thermodynamic limit:

$$PES \approx E^{\#} - \lim_{t\to\infty}\left[\lim_{\delta\to 0} V(Q,t)\right]. \quad (1.8)$$

To map out the topology of $\Omega/\sim$, that is, the connectivity of the basins of attraction of destiny points, it is necessary to input a dedicated autoencoder (AE2) with time series resulting from metadynamics trajectories at time resolution $\Delta t \propto \delta$. Taking into account that the basins B_j are

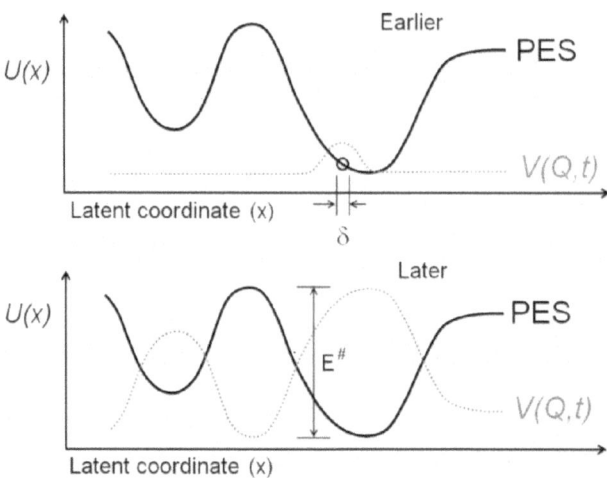

FIGURE 1.7 Pseudo-ergodic metadynamics reconstruction of a metamodel represented by modulo-basin transitions.

denumerable for a compact latent manifold ($\delta \to 0$), the data for a single metadynamics trajectory is organized in the matrix $\boldsymbol{M} = \left[\, \boldsymbol{b}(0)\boldsymbol{b}(1)...\right]$, where the binary vector $\boldsymbol{b}(n) \equiv \boldsymbol{b}(n\Delta t)$, $n = 0, 1, ...$ is given by $b_j(n) = 1$ if the system is allocated to basin B_j at time $n\Delta t$, and $b_j(n) = 0$, otherwise. To fix notation, let $j, j', k, k',...$ index basins in the latent manifold, then if a basin transition $B_j(n) \to B_{j'}(n + 1)$ materializes in the inputted $\boldsymbol{M}$, AE2 provides output signaling the existence of a saddle $S_{jj'}$, so that a steepest descent pathway connects B_j and $B_{j'}$. Figure 1.8 shows the topological description of the quotient space as a disconnectivity graph [25] obtained through pseudo-ergodic metadynamics and tensorially incorporated to the metamodel for the system described in Figure 1.5.

As AE2 is trained with a series of metadynamics matrices $\boldsymbol{M}_m$ ($m = 1$, 2,...) generated by different metadynamics runs, a basin transition $k \leftrightarrow k'$ together with at least one of the spillover basin transitions $j \leftrightarrow k$, $j \leftrightarrow k'$, $j' \leftrightarrow k$, $j' \leftrightarrow k'$ may materialize. In that case, the saddle $S_{jj'}$ gets "tensorially revised" to a higher index saddle: $S_{jj' \cdot kk'} = S_{jj'} \otimes S_{kk'}$, which means that two of the Hessian eigenvalues at the saddle point $S_{jj' \cdot kk'}$ containing the lines of steepest descent joining respectively basins B_j, $B_{j'}$ and B_k, $B_{k'}$ are negative, in contrast with the original model ($S_{jj'}$), where only one negative eigenvalue was inferred. In other words, the saddle originally assumed by the autoencoder to be of index 1 proved to be of index 2 as the metadynamics progressed. Of course, saddles of index higher than two are inferred by the

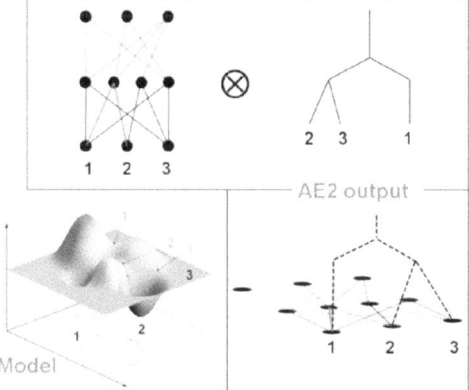

FIGURE 1.8   Metadynamics reconstruction of the topological metamodel for the keto-enol isomerization of acetone.

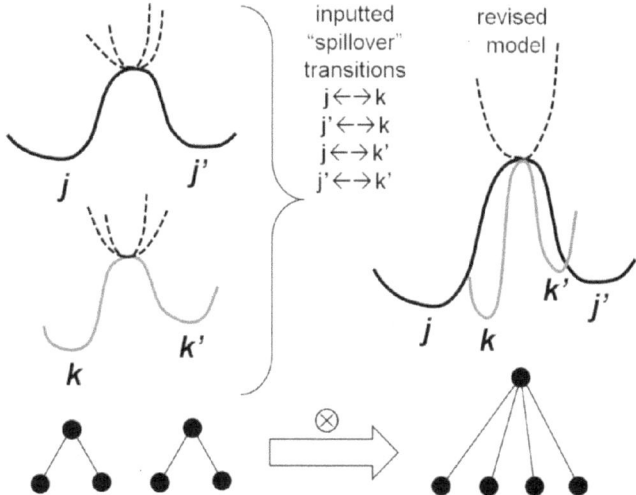

FIGURE 1.9   Schematics of the updating of topological relations between basins of attraction that takes place as the metadynamics progresses, imposing an iterative updating of the metamodel.

autoencoder in a similar fashion through progressive model revision as more metadynamics matrices are inputted to train the network, and more spillover basin transitions are detected. Figure 1.9 depicts schematically a step in the progressive revision of the metadynamics mapping of the $\Omega/\sim$ topology that represents the connectivity of the basins of attraction of destiny points.

1.10 METAMODELS FOR THE DIGITAL MIND

As previously discussed, given an observable process that generates data organized as a time series, a central goal of modern science is to discover a physical model cast in the intrinsic coordinates of the system. When represented by differential equations in the latent manifold, the model is subject to the constraint that it must be decodable into the dynamical system that underlies the process. This has traditionally been an overarching goal in data-driven model discovery for dynamical systems. However, data science practitioners reckon that the level of multiscale complexity of the system often precludes the discovery of models cast in the terms previously described. Parsimonious geometric models often elude discovery, especially when dealing with the vast arrays of biological, biomedical, or cosmological data, as illustrated in this book.

Those systems call for new methods of data-driven model discovery, with few such approaches currently underway. This book portends to spearhead such efforts in a systematic comprehensive way. Most significantly, the topological metamodel emerges as a distinctive possibility, as it introduces drastic simplifications to encode the dynamical system as a graph that retains and articulates all relevant topological features of the vector field. Furthermore, the dynamics turns into a pattern on the graph that can be trained and ultimately decoded into the full coordinate framework, fleshing out the dynamical system, as described in the subsequent chapters.

As shown in Chapter 2, metamodels of a dynamical system can in fact be taxonomically organized in a hierarchy, and can be concatenated through a hidden Markov model that retrieves the information from the highest level of abstraction and decodes it at the most concrete (geometric) level. Thus, if we assume the canonical projection $\pi : \Omega \to \Omega/\sim$ given by $\pi(x) = \omega(x)$, we can see that the model on Ω and the metamodel on $\Omega/\sim$ are related by the hidden Markov model $P(x, t) = \sum_{\omega} P(x|\omega) P(\omega, t) = P(x|\omega(x)) P(\omega(x), t)$, where $P(\omega, t)$ is the probability that the basin of attraction of the destiny set ω is realized in the metamodel at time t, and $P(x|\omega)$ is the conditional probability of state x given that the system is in the basin of attraction of ω when described at the higher level of abstraction given by the metamodel. In fact, for a hierarchy of metamodels (Chapter 2), we may define a hierarchy of tandem hidden Markov models that pairwise links them along the taxonomy as we progressively increase the level of abstraction.

The topological metamodel presupposes an adiabatic limit, that is, the coarse-graining of time associated with it is such that equilibration has been reached within the basins of attraction of the critical points or sets of points that represent destiny states. This implies that the conditional probability density $p(x|\omega)$ is given by the Boltzmann weight:

$$p\left(x|\omega\right) = \delta_{\omega,\omega(x)}\left[e^{-\beta U(x)} \bigg/ \int_{B(\omega)} e^{-\beta U(y)} dy\right], \qquad (1.9)$$

where $\delta_{\omega,\omega(x)}$ is Kronecker's delta, $\beta = (k_B T)^{-1}$ with k_B = Boltzmann constant and T = absolute temperature, $B(\omega)$ is the basin of attraction of ω, and the integration extends over the entire basin.

In his plan to create a digital mind mimicking the workings of the human brain, Ray Kurzweil persuasively argues that the brain contains hierarchies of general pattern recognizers, of which the neocortex contains approximately 300 million, accounting for most aspects of sensing, learning, and thinking [26]. This picture should be understood within Kurzweil's intent to expand the neocortex beyond its skull enclosure. The hierarchical organization of pattern recognizers at increasingly higher levels of abstraction lends itself to a model of the mind where patterns are linked via hidden Markov processes. Thus, the concept of topological metamodel fits naturally in Kurzweil's model for a digital mind in relation to conceptualizing and analyzing time series data absorbed at different levels within a conceptual hierarchy of pattern recognizers. In such mental operations, patterns that represent dynamical processes, or, more generally, events, are enshrined in metamodels that represent higher levels of abstraction. Furthermore, geometric model and its ensuing hierarchically organized topological metamodels are pairwise linked by hidden Markov processes, as in Kurzweil's plan for the digital mind.

If in fact a neocortex-expanding AI is constructed following Kurzweil's blueprint for the digital mind, then most probably the topological metamodels implemented in this book will enable the AI-empowered discovery of physical laws underlying dynamical systems.

REFERENCES

1. Russell S, Norvig P (2020) *Artificial intelligence: A modern approach*. Pearson, London, UK.
2. Kelleher JD (2019) *Deep learning*. The MIT Press, Cambridge, MA.
3. Chollet F (2019) *Deep learning with python*. Manning Press, Shelter Island, New York.

4. Atienza R (2020) *Advanced deep learning with TensorFlow 2 and Keras*, 2nd Edition. Packt Publishing, Birmingham, UK.

5. Schmidhuber J (2015) Deep Learning in Neural Networks: An Overview. *Neural Netw* 61:85–117.

6. Fernández A (2016) *Physics at the biomolecular interface.* Springer International Publishing, Switzerland.

7. Aliper A, Plis S, Artemov A, Ulloa A, Mamoshina P, Zhavoronkov A (2016) Deep Learning Applications for Predicting Pharmacological Properties of Drugs and Drug Repurposing Using Transcriptomic Data. *Mol Pharm (ACS)* 13:2524–2530.

8. Lavecchia A (2019) Deep Learning in Drug Discovery: Opportunities, Challenges and Future Prospects. *Drug Disc Today* 24:2017–2032.

9. Jiménez J, Škalič M, Martínez-Rosell G, De Fabritiis, G (2018) Kdeep: Protein–Ligand Absolute Binding Affinity Prediction via 3D-Convolutional Neural Networks. *J Chem Inf Mod* 58:287–296.

10. Stepniewska-Dziubinska MM, Zielenkiewicz P, Siedlecki P (2018) Development and Evaluation of a Deep Learning Model for Protein–Ligand Binding Affinity Prediction. *Bioinformatics* 34:3666–3674.

11. Hassan-Harrirou H, Zhang C, Lemmin T (2020) RosENet: Improving Binding Affinity Prediction by Leveraging Molecular Mechanics Energies with an Ensemble of 3D Convolutional Neural Networks. *J Chem Inf Model* 60:2791–2802.

12. Fernández A (2020) Artificial Intelligence Teaches Drugs to Target Proteins by Tackling the Induced Folding Problem. *Mol Pharm (ACS)* 17:2761–2767.

13. Callaway E (2020) 'It Will Change Everything': DeepMind's AI Makes Gigantic Leap in Solving Protein Structures. *Nature* 588(7837):203–205. doi: https://doi.org/10.1038/d41586-020-03348-4

14. Heaven WD (2020) DeepMind's Protein-Folding AI Has Solved a 50-Year-Old Grand Challenge of Biology. *MIT Tech Rev.* Nov 30, 2020. https://www.technologyreview.com/2020/11/30/1012712/deepmind-protein-folding-ai-solved-biology-science-drugs-disease/

15. Brunton SL, Kutz NJ (2019) *Data-driven science and engineering: Machine learning, dynamical systems and control.* Cambridge University Press, UK.

16. Tiumentsev Y, Egorchev M (2019) *Neural network modeling and identification of dynamical systems.* Academic Press, San Diego, USA.

17. Champion K, Lusch B, Kutz JN, Brunton SL (2019) Data-Driven Discovery of Coordinates and Governing Equations. *Proc Natl Acad Sci USA* 116:22445–22451.

18. Schmidt M, Lipson H (2009) Distilling Free-Form Natural Laws from Experimental Data. *Science* 324:81–85.

19. Fernández A (1985) Center-Manifold Extension of the Adiabatic-Elimination Method. *Phys Rev A* 32:3070–3076.

20. Born M, Oppenheimer JR (1927) Zur Quantentheorie der Molekeln. *Ann Physik* 389:457–484.

21. Fernández A (2014) Chemical Functionality of Interfacial Water Enveloping Nanoscale Structural Defects in Proteins. *J Chem Phys* 140:221102.

22. Fernández A (2021) *Artificial intelligence platform for molecular targeted therapy: A translational approach.* Chapter 9. World Scientific Publishing, Singapore.

23. Fernández A (2020) Deep Learning Unravels a Dynamic Hierarchy While Empowering Molecular Dynamics Simulations. *Ann Physik (Berlin)* 532:1900526.

24. Bussi G, Laio A (2020) Using Metadynamics to Explore Complex Free-Energy Landscapes. *Nature Rev Phys* 2:200–212.

25. Krivov S, Karplus M (2002) Free Energy Disconnectivity Graphs: Application to Peptide Models. *J Chem Phys* 117:10894.

26. Kurzweil R (2012) *How to create a mind.* Penguin, New York.

## CHAPTER 2

Topological Methods for Metamodel Discovery with Artificial Intelligence

*Nevertheless, I suspect he was
not quite capable of thinking.
Thinking is neglecting differences,
generalizing, abstracting.
In Funes' cluttered world there was
nothing but details, immediacy.*

– Jorge Luis Borges: Funes The Memorious
(translation by the author)

2.1 AI-BASED METAMODEL DISCOVERY FOR HIERARCHICALLY COMPLEX DYNAMICAL SYSTEMS

With the leveraging of artificial intelligence (AI) [1] and, in particular, machine learning approaches [2, 3], dynamical systems have found a new fertile ground for further development [4, 5]. Showcase problems in applied mathematics, including the Lorenz strange attractor, reaction-diffusion spatio-temporal systems, and fluid dynamic flows captured by

DOI: 10.1201/9781003333012-3

the Navier–Stokes equation are being examined in a new guise as auto-encoders identify parsimonious models with reduced dimensionality [6]. Such models are meant to provide the physical underpinnings of the phenomena enshrined in time series data or generated by raw differential equations.

Machine learning, or, more broadly, AI, is being leveraged for model discovery of dynamical systems underlying data represented as a time series. The data regression system, which in this context is a neural network predicting future behavior, is trained by the time series and regarded as the model itself. However, this model is heavily parametrized and hence too "fragile" to allow for extrapolation [4]. In other words, such models are not really amenable to yield physical laws, the way other data-regression approaches are [7, 8]. This statement has been voiced repeatedly and hints at some level of dissatisfaction: machine learning and AI in general are very efficacious at providing predictive models when trained on a sufficiently long time series but often do a poor job at providing physical insights regarding the underlying dynamical system.

This problem gets significantly amplified as we turn to biological or, more broadly, biomedical matter [9]. It is widely felt that, when examined in their multiscale richness and complex heterogeneity, dynamical systems in biology or biomedicine cannot reach the level of maturation required to be subsumed into applied mathematics. This statement should be interpreted in the sense that we lack sparse enough models that provide physical underpinnings of biological/biomedical phenomena and are suitable for extrapolation. Will the leveraging of alternative AI-based approaches change the status quo? This chapter portends to address this problem and provide insights that will be methodologically fleshed out in the subsequent chapters to enable metamodel discovery by reverse engineering time series stemming from highly complex multiscale realities.

A key question in fostering the mathematical maturation of model discovery in biology may be cast as follows: What constitutes a parsimonious model that provides physical underpinnings of biological phenomena? A standard answer with broad bearing on most problems considered tractable in applied mathematics is: "a sparse system of differential equations on a smooth manifold of latent coordinates" [6, 8]. As this chapter argues, this may be simply too much to ask for in the context of biological matter. Furthermore, this is not necessarily the format or framework that AI would typically choose for model discovery, given the "dimensionality curse" associated with the molecular reality of biological systems [9]. In principle, a single autoencoder that optimizes for sparsity in the discovery of the latent manifold might not provide a satisfactory solution to the modeling problem. Molecular reality *in vivo* typically has well over a million coordinates required to specify the state of the system, and the extent of connectivity parametrization for an autoencoder capable of handling such level of complexity would be simply enormous, implying that the training and variational optimization of the neural network would be off limits, at least with current computational capabilities [10].

This chapter squarely addresses this matter. To do so, it leverages AI methods to circumvent the difficulties associated with model discovery for time-dependent phenomena arising in soft or biological matter. In essence, as we shall show, AI dwells on a paradigm of what constitutes a parsimonious metamodel that is significantly different from the one adopted by applied mathematicians [9]. Thus, to identify the most economic yet faithful metamodel, AI will be shown to use two or more tandem autoencoders instead of one, as it is typically done in model discovery [6]. The autoencoders are coupled and become fully compatible with each other at the completion of the parameter optimization process, as defined precisely in the subsequent sections. In the simpler cases where two autoencoders are required, the second autoencoder translates the dynamic information embossed in the latent manifold, turning it into a topological dynamics metamodel [11] which can be decoded, and enables significant propagation of the dynamics into the future. This property is essential for coverage of realistic timescales relevant to the level of state extrapolation required. Crucially, the topological dynamics metamodel constructed by AI is essentially a pattern-based model, not a system of differential equations, as it would be expected for standard model discovery in dynamical systems. This does not mean that AI is discovering laws without equations, but simply that AI adopts a different way of casting models susceptible of extrapolation as recognizable physics laws. Thus, AI may not straightforwardly give us the equations that govern *in vivo* protein folding, but the underlying physics discovered may in all likelihood be cast in terms of AI-interpretable topological patterns that signal the commitment of the chain to fold into a steady conformation [12].

These topological dynamic methods are likely to advance the field of AI-based model discovery as they enable the reverse engineering of time series data stemming from vastly complex hierarchical realities. Such contexts are illustrated, for example, by the cellular setting that assists and expedites molecular processes, which have been hitherto considered off limits to machine learning approaches to dynamical systems.

At this stage of development, the topological methods introduced yield AI-recognizable patterns but do not beget latent differential equations that have traditionally been the hallmarks of model discovery. This may be viewed as a limitation in some sense, but we argue that that assertion is perhaps a reflection of narrow-mindedness. Synergistic efforts involving AI are likely to dominate future human endeavor in science, and AI systems are very much attuned to encode and process patterns in metamodels such as those produced by the topological approaches introduced in this book.

The impact of the topological methodology is likely to be broad, as it would render tractable problems in dynamical model discovery that have been hitherto considered off-limits by applied mathematicians that are currently incorporating machine learning in their toolbox. Thus, ultra-complex realities recreating cellular environments that influence and steer molecular dynamics, such as those treated in Chapters 3 and 4, are likely to be within reach as topological methods are incorporated to the AI-empowered metamodel discovery.

These advances represent substantial contributions to biomedical research and have practical implications for molecular targeted therapy, where the assessment of the *in vivo* reality is crucial to discover novel therapeutic agents capable of efficacious and highly specific molecular intervention, as shown in Chapter 4.

2.2 AUTOENCODERS OF LATENT COORDINATES IN DYNAMICAL SYSTEMS

Since the dawn of modern science, western civilization upheld the belief that understanding the workings of the universe pivots on finding fundamental equations that govern physical processes. In the case of a dynamical system, the underlying differential equations represent the basic model assumed to enshrine the physical laws that underpin the process described. The specific constraints, conservation principles, and symmetries of the system must all be taken into account when positing the differential equations. After centuries of work within this paradigm, it would be interesting to see how the leveraging of AI in synergy with the human endeavor will affect the choice of the format within which the physical laws are encoded and how this choice impacts the field of dynamic modeling. With the exponential development of computer technologies [10], we may soon be witnessing a paradigm revision. Equation-based modeling may or may not remain the dominant paradigm. Other types of dynamic descriptors are already making strides and contributing to a new understanding of the universe or at least of the dynamic multiscale complexities of reality [9]. These descriptors are certainly different and arguably more pliable than what biological humans managed to achieve so far.

The discovery of physical laws distilled from sequential data representing a time series remains a major challenge as well as an imperative to enhance our understanding and control of physical processes. In recent times, this type of data-driven model discovery has been fuelled by significant breakthroughs [7–9, 13]. Yet, we also live in an era of big data, especially stemming from fields like biology and astrophysics, where the wanton multiscale complexity of the data organization makes model discovery particularly daunting. It is unclear whether the enormous richness of the data describing *in vivo* contexts at the molecular level is amenable to the kind of parsimonious models based on differential equations that humankind holds dear and has sought since time immemorial. More than in any other field, in biology it is likely that a willy-nilly application of Occam's razor may lead to self-inflicted wounds.

Deep learning (DL) approaches realized through autoencoder architectures have proven particularly valuable for the discovery of data-driven models represented by differential equations framed on "essential coordinates." The latter, often referred to as "latent coordinates" [8], span the so-called center manifold in dynamical systems [9, 14]. This reduction entails a significant dimensionality reduction, and usually identifies the enslaving "slow" process that dynamically subordinates fast-relaxing modes [4, 14] and serves to encode the physical process we seek to model. Latent coordinates need to be selected very carefully, by weighing the extent of not only dimensionality reduction and the compactness and smoothness of the latent manifold but also the economy of the resulting model. This economy is typically assessed by the complexity of the differential equations in latent space, quantified by the number of nonlinear terms [6].

In generic terms, an autoencoder seeks to identify latent coordinates $\mathbf{x}$ as output of a feed-forward neural network (NN) with multiple hidden layers. The NN is inputted observable vectors, denoted generically as $\mathbf{z}$, that serve to train the network, as the decoding of $\mathbf{z}$ from the latent coordinates must reproduce the inputted $\mathbf{z}$-value (Figure 2.1). The autoencoder is optimized variationally, meaning that the activation weight parametrization of the multilayered encoding and decoding functions, denoted respectively as

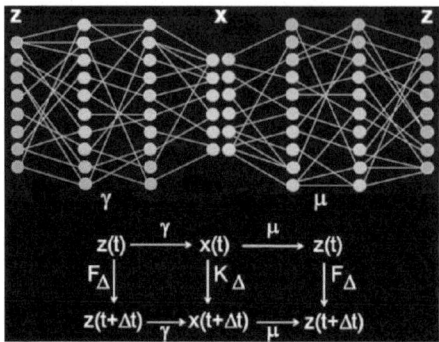

FIGURE 2.1 Generic scheme of the neural network (NN) architecture for autoencoder (γ, μ). The vector $z \in W$ represents the state of the system, and $x \in \Omega$ represents an encoded latent state that is decoded back into z. The encoding process entails a dimensionality reduction with $dim\Omega < dimW$ and, concomitantly, a coarse-graining of time in multiples of a time step Δ associated with propagation of the dynamics on the latent manifold Ω. Gray disks represent nodes in hidden layers for encoder $\gamma = \gamma(\Theta)$ and decoder $\mu = \mu(\Theta)$, where Θ denotes the weights of node connections that are variationally optimized according to a loss function $\mathcal{L} = \mathcal{L}(\Theta)$. The weights realizing the variational mimimum $\left(\Theta = arg\ min\ \mathcal{L}\right)$ in the training of the NN are optimal at making the diagram commutative.

γ and μ, minimizes the loss function $\mathcal{L}(\gamma,\mu)$ that measures the efficacy in the recovery of the input vectors:

$$\mathcal{L}(\gamma,\mu) = Q^{-1}\sum_{z\in\mathfrak{F}}\|z-(\mu\circ\gamma)z\|^2; \gamma,\mu = \arg\min_{\gamma,\mu}\mathcal{L} \qquad (2.1)$$

where $\mathfrak{F}$ is the training set and $Q = |\mathfrak{F}|$ is its cardinal, for now regarded as a hyperparameter fixed to avoid overfitting [2–5] in accord with the dimensions of the NN.

The autoencoder is functionally operative to model the dynamical system that underlies the given time series $\{z(t_0 + n\Delta t), z(t_0 + (n + 1)\ \Delta t)\}_{n = 0, 1, 2, ..., L}$ if and only if the following relations hold for $t = t_0 + n\Delta\ t$ with $n = 0, 1, 2, ..., L$ (L>>1) and any initial time t_0:

$$(K_\Delta\circ\gamma)z(t) = \gamma z(t+\Delta t) = (\gamma\circ F_\Delta)(F_\Delta\circ\mu\circ\gamma)z(t);$$
$$z(t) = (\mu\circ K_\Delta\circ\gamma)z(t) = z(t+\Delta t) \qquad (2.2)$$

where F_Δ, K_Δ are the infinitesimal time maps in state space W and latent manifold Ω, respectively (cf. Figure 2.1). In other words, as the autoencoder becomes variationally optimized, the functional relations $K_\Delta \circ \gamma = \gamma \circ F_\Delta$; $F_\Delta \circ \mu = \mu \circ K_\Delta$ hold.

Thus, diagram commutativity (Figure 2.1) becomes the key property that enables us to assert that the choice of latent coordinates $x \in \Omega$ was the "right" one in the sense that it captures the entrainment of the dynamics on W by the reduced dynamics on the latent manifold Ω. The commutativity rules ensuring the compatibility of raw and latent dynamics may be cast in terms of derivatives using the chain rule [6] as Δt is taken to the infinitesimal limit dt.

The importance of identifying the "right" latent coordinate frame that yields a parsimonious model of the physical process cannot be overemphasized [4], and merits a historical digression. It suffices to recall the extremely awkward description of apparent planetary motion, with circles within circles, put forth by Claudius Ptolemy (circa 100–170 AD), who "incorrectly" placed the earth at the center of coordinates. This is actually a rather unkind portrayal of a major intellectual achievement. After all, the Ptolemaic system proved highly accurate in spite of its laboriousness as a predictive tool for celestial motion. Be as it may, it took well over 1400 years to replace the ultra-complicated geocentric system of Ptolemy for the far more straightforward heliocentric coordinate system, a veritable intellectual feat and an incredibly bold step by Nicolaus Copernicus (1473–1543). The "right coordinates" enabled proper regression of the astronomic observational data by Johannes Kepler (1571–1630), which in turn enabled the towering discovery of the gravitational model of universal validity by Isaac Newton (1643–1727) (Figure 2.2).

2.3 DEEP LEARNING SCHEME TO DISCOVER UNDERLYING DIFFERENTIAL EQUATIONS FROM TIME SERIES

Model discovery in dynamical systems has been maturing for some time, to the extent that it has been integrated to the corpus of applied mathematics. Furthermore, many applied mathematicians have incorporated DL to their toolbox as they make forays in reverse engineering of dynamical systems [4]. The overarching goal is to develop regression methods that enable extraction of parsimonious dynamics from large data organized as time series $\{z(t)\}_t$ with t given as multiples of a fixed time step [4, 6]. With the advent of DL, autoencoder architectures have successfully identified latent (essential) coordinate frames with significant dimensionality reduction ($\dim\Omega < \dim W$). Ultimately, the parsimonious description should become solvable in the form of a differential system $\dot{x} = K(x)$,

FIGURE 2.2   Four key players in the discovery of a universal model for celestial dynamics. Clockwise from top left: Claudius Ptolemy (c. 100–170 AD), Nicolaus Copernicus (1473–1543), Johannes Kepler (1571–1630), and Isaac Newton (1643–1727). Ptolemy proposed the first latent coordinate system (geocentric) and developed a predictive scheme, although the latent coordinate frame he chose was suboptimal and hence introduced huge operational complications; Copernicus identified the right latent coordinate frame (heliocentric) leading to a parsimonious description; using the Copernicus framework, Kepler was able to perform the data regression that yielded an empirical heuristic model; and Newton created a sparse model that subsumed Kepler's data regression scheme and enabled extrapolation, establishing a law of wide applicability.

which spans a differential system $\dot{z} = F(z)$ for the time series vector z. Yet, a mere stark reduction in dimensionality does not necessarily ensure a parsimonious description [6]. The variational optimization of the auto-encoder should simultaneously consider both the dimensionality of the coordinate frame for the latent manifold and the sparsity of the equations

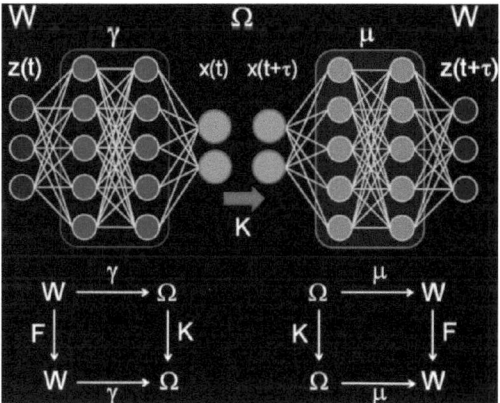

FIGURE 2.3 Scheme of autoencoder for latent dynamics optimizing simultaneously the dimensionality reduction and the parsimony of latent dynamical system governed by the flow K. The optimization is extended further relative to the architecture displayed in Figure 2.1 by adding the extra term $\mathcal{L}_s(K)$ to yield the loss functional $\mathcal{L} = \mathcal{L}(\Theta, K) = \mathcal{L}(\Theta) + \mathcal{L}_s(K)$. The optimal Θ, K pair realizing the variational minimum is precisely the one that makes the bottom diagrams commutative.

in the latent manifold, with the latter measured by the number of terms in the flow (vector field) that determines $\dot{z}$

The autoencoder architecture and the required dual diagram commutativity between the latent flow K and the given time series flow F are jointly represented in Figure 2.3. We adopt a time step τ as the time series hyperparameter to discretize the flows by coarse-graining time resolution. The loss function $\mathcal{L}(\gamma, \mu, K)$ now includes five terms: $\mathcal{L}_r(\gamma, \mu)$, previously introduced, weights the efficacy of the recovery of the z-value upon encoding and decoding, $\mathcal{L}_s(K)$ accounts for the sparsity of the latent flow K, and the terms $\mathcal{L}_C(\gamma, K), \mathcal{L}_C(\gamma, \mu), \mathcal{L}_C(\gamma, \mu, K)$ represent the penalties associated with imperfect diagram commutativity that need to be imposed to guarantee the dynamic compatibility of the z and x descriptions of the system. Thus, we get

$$\mathcal{L}(\gamma, \mu, K) = \mathcal{L}_r(\gamma, \mu) + \vartheta_s \mathcal{L}_s(K) + \vartheta_C[\mathcal{L}_C(\gamma, K) + \mathcal{L}_C(\gamma, \mu) + \mathcal{L}_C(\gamma, \mu, K)],$$

(2.3)

where the relative weights ϑ_s, ϑ_C are hyperparameters and

$$\mathcal{L}_r(\gamma, \mu) = Q^{-1} \sum_{z \in \mathfrak{F}} ||z - (\mu \circ \gamma)z||^2,$$

(2.4)

$$\mathcal{L}_C(\gamma, K) = Q^{-1} \sum_{z \in \mathfrak{F}} \|(K \circ \gamma)z - (\gamma \circ F)z\|^2, \qquad (2.5)$$

$$\mathcal{L}_C(\gamma, \mu) = Q^{-1} \sum_{z \in \mathfrak{F}} \|(F \circ \mu \circ \gamma)z - (\mu \circ \gamma \circ F)z\|^2, \qquad (2.6)$$

$$\mathcal{L}_C(\gamma, \mu, K) = Q^{-1} \sum_{z \in \mathfrak{F}} \|(\mu \circ K \circ \gamma)z - (F \circ \mu \circ \gamma)z\|^2, \qquad (2.7)$$

with $Q = |\mathfrak{F}|$ indicating the cardinal of the training set.

To determine the sparsity term $\mathcal{L}_s(K)$, we first take the limit case where τ becomes the time infinitesimal dt and assume that $\dot{x} = K(x) = AP(x)$, where $A = [Aij]$ is an mxp matrix ($m = \dim\Omega$) whose coefficients are variationally optimized jointly with the parametrization of the autoencoder maps γ and μ, and $P(x)$ is a p-vector given by $P(x)^T = \left(1\, x_1\, x_2 x_3\, x_1^2\, x_2^2\, x_3^2\, x_1 x_2\, x_2 x_3\, x_1 x_3\, x_1 x_2 x_3 \ldots\right)$ consisting of p functional terms. The terms do not need be of polynomial form, and they may be chosen in accord with the type of symbolic regression. Thus, the variational optimization of A is tantamount to a symbolic regression in the latent manifold Ω. To ensure the sparsity of the model, we define the loss term as $\mathcal{L}_s(K) = \|A\|_F$, where $\|.\|_F$ is the Frobenius norm given by

$$\|A\|_F = \left[\sum_{i=1}^{m} \sum_{j=1}^{p} A_{ij}^2 \right]^{1/2} \qquad (2.8)$$

The incorporation of this term into the loss function given by Equation (2.3) ensures that the latent manifold is selected to provide a parsimonious model with reduced dimensionality governed by the simplest possible set of differential equations in consonance with the time series provided. A different but equivalent formulation of the autoencoder has been given elsewhere [6], and uses the chain rule of derivation rather than diagram commutativity to infer the proper loss functional.

2.4 AUTOENCODERS FOR MOLECULAR DYNAMICS OF BIOLOGICAL MATTER

As applied mathematicians make their forays combining dynamical systems with machine learning, they adopt specific systems that have traditionally represented showcases for model building. Examples of such systems are the

Lorenz attractor generated by three coupled ordinary differential equations, the spatio-temporal reaction-diffusion systems, and the hydrodynamics and turbulence models governed by the Navier–Stokes equation [6, 13]. Seldom, if ever, we had a chance to evaluate how autoencoder techniques pan out in the realm of molecular dynamics for ultra-complex multiscale many-body systems. We are specifically referring to the discovery of models that distill the cooperative collective motion of ensembles of atoms, atomic groups, molecules, or molecular assemblages in condensed phases characterized as clusters, liquids, glasses, crystals, polymers, and so on, under overarching categories such as soft matter and biological matter. The foundational approach to model discovery for such systems has been traditionally provided by statistical mechanics. These methods have met considerable success, except in the realm of biological matter, where problems such as the discovery of protein folding pathways, or the role of *in vivo*/cellular contexts in expediting the folding process cannot be even remotely addressed due to their sheer complexity and heterogeneity. Interacting molecular units are simply too diverse, and the cellular environment is too complex at multiple scales and heterogeneous for statistical mechanics to make successful forays in biology [9, 15, 16]. This is precisely the context where AI may empower dynamical systems by leveraging highly specialized autoencoders and even batteries of autoencoders, as described subsequently.

To furnish a general framework, we may start by noting that we are dealing with N particles that may be free or tethered through covalent linkages forming assemblages such as biopolymers that may interact with one other through ephemeral or permanent associations that do not involve covalent bonds. A protein chain embedded in an aqueous environment and interacting with other biomolecular entities such as protein enzymes, chaperones, or ribosomes represents the quintessential situation that we wish to address in this book as we learn to leverage AI methods and integrate them on a model discovery platform.

Keeping for now the discussion at its most generic level, let us consider an ensemble of N particles that has associated with it an internal energy $\mathcal{U} = \mathcal{U}(\hat{z}), \hat{z} \in \mathbb{R}^{3N}$ only dependent on interparticle distances and hence invariant upon isometries – distance-preserving maps – of $\mathbb{R}^3$. Then, the state or configuration of the system may be represented as a point z in the quotient space $W = \mathbb{R}^{3N}/E^+(3)$, where $E^+(3)$ is the special Euclidean group of isometries of $\mathbb{R}^3$ including only rigid-body translations and rotations but excluding reflections [17]. The latter are excluded because they do not preserve the chirality (handedness) of asymmetric tetrahedral carbon groups, which constitutes a key constraint: chirality is known to be strictly preserved in biology.

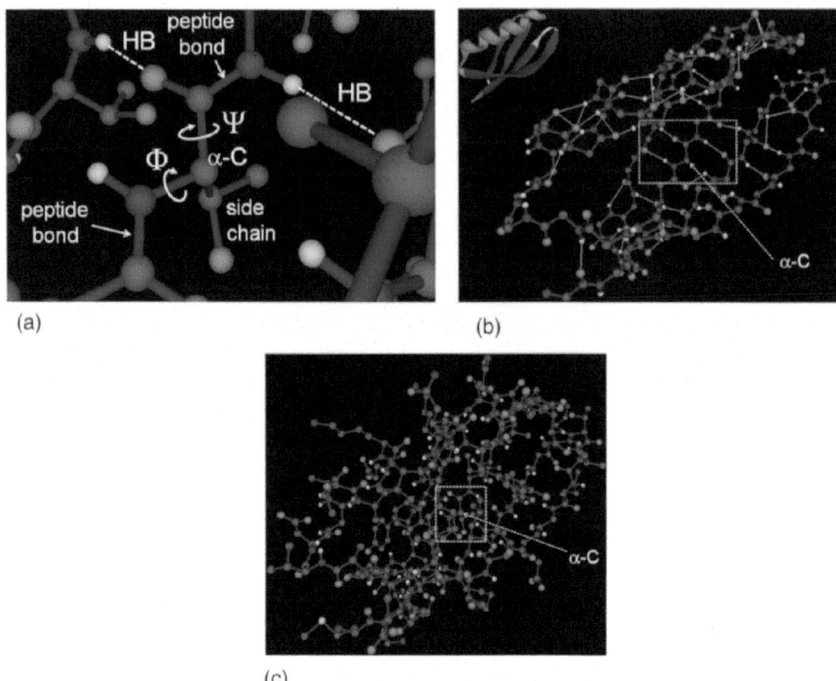

(a)

(b)

(c)

FIGURE 2.4 Latent coordinates for a folding protein chain. **(a)** Individual residue unit within a protein chain attached to an adjacent residue through a torsionally rigid linkage known as peptide bond. For the sake of illustration, a particular protein has been selected, namely the thermophilic variant of the B1 domain of protein G from *Streptococcus*, whose native 3D-structure is found at the entry 1GB4 in the protein data bank (PDB). In the latent coordinate frame, the state of each residue is represented by a pair of coordinates (Φ, Ψ) representing the torsional dihedral degrees of freedom of the protein backbone. The amino acid type for the residue, describing the local chemical composition of the chain, is identified by the side-chain group that is covalently linked to the alpha-carbon in the protein backbone. In turn, polar groups in the backbone and side chain can be engaged in orientationally and distance-dependent noncovalent linkages known as hydrogen bonds (HB, dashed lines), whose cumulative folding-stabilizing effect becomes a determinant of the protein 3D-structure. **(b)** Zoom out of the detail shown in (a), revealing the backbone HB pattern and topology (ribbon rendering) of the native structure. Backbone HBs exposed to water are indicated by thin green lines (buried ones in gray), and they signal structural deficiencies, since the structure may get locally disrupted when the polar groups paired by intramolecular HBs get hydrated, that is, interact with surrounding molecules of the aqueous solvent. **(c)** All-atom rendering of the protein folded into its native structure reported in PDB.1GB4.

Thus, W is (3N-6)-dimensional, and its coordinates represent all the internal degrees of freedom of the system. In addition to the reduction modulo isometries, another quotient is required to represent the latent manifold Ω. This further dimensionality reduction depends on the constraints brought about by the covalent linkages that tether specific units in the system. For example, if we are studying the dynamics of protein folding, we note that high-frequency vibrational motions involving covalently paired atoms may be averaged out as their associated timescales, in the femtosecond (fs) range [18], are incommensurably shorter than those associated with soft modes represented by dihedral torsions of the polymer backbone (Figure 2.4), typically in the nanosecond (ns) or sub-ns range. A similar reduction applies to planar angular vibrations, with frequencies in the order of $(fs)^{-1}$ to $(10fs)^{-1}$ [9, 18]. These simplifications, point to a parsimonious model representing an adiabatic system that incorporates only soft modes of the chain. Thus, as protein folding is represented as a dynamical system with the protein chain searching in conformation space in an *in vitro* setting, Ω becomes a compact manifold in the form of a 2M-torus, where M is the number of amino acid units in the chain [12] (Figure 2.4). This latent manifold is deduced assuming that the water molecules surrounding the protein chain that explores conformation space are treated implicitly, that is, their influence is energetically subsumed in the potential energy function U only dependent on distances between the protein chain subunits and the local configurations that result thereof.

The dynamics in W obeys the basic laws of physics:

$$\dot{z} = v; \dot{v} = -\nabla_z U \qquad (2.9)$$

To properly define $U=U(\mathbf{z})$, we introduce the following:

Theorem 2.1

There exists a map $U : \mathbb{R}^{3N}/E^+(3) \to \mathbb{R}$ that makes the following diagram commutative:

$$\mathcal{U}$$

$$\mathbb{R}^{3N} \to \mathbb{R}$$

$$\pi \downarrow \quad \nearrow U \qquad \Rightarrow \quad U \circ \pi = \mathcal{U} \qquad (2.10)$$

$$\mathbb{R}^{3N}/E^+(3)$$

where $\pi : \mathbb{R}^{3N} \to \mathbb{R}^{3N}/E^+(3)$ is the canonical projection associating each point in $\mathbb{R}^{3N}$, specifying the 3D-coordinates of each of the N particles of the system, with the set of points in $\mathbb{R}^{3N}$ resulting from rotations and translations of the N-particle system treated as a rigid body, that is pre-serving all interparticle distances. Thus, the projection associates a state of the system with its class in quotient space, that is, with the collection of all points contained in the group orbit generated by the action of the Euclidean group on the point in the domain.

To prove this "factorization" result, it suffices to note that the potential energy $\mathcal{U} : \mathbb{R}^{3N} \to \mathbb{R}, \mathcal{U} = \mathcal{U}(\hat{z}), \hat{z} \epsilon \mathbb{R}^{3N}$ is invariant on the orbits of $E^+(3)$ in $\mathbb{R}^{3N}$.

An autoencoder (γ, μ) yielding a parsimonious model in Ω would require that the autoencoder constructs $\tilde{U}(x) = U(\mu(x))$, so the dynamic equations in the latent manifold become

$$\dot{x} = \tilde{v}; \dot{\tilde{v}} = -\nabla_x \tilde{U}(x). \tag{2.11}$$

This implies that in the infinitesimal limit $\tau = dt$, the potential energy function may be obtained by noting that the sparse map K defined by the autoencoder obeys

$$\dot{x}(t) = K(x(t)) = -\int_0^t \nabla_x \tilde{U}(x(t'))dt' + \dot{x}(0). \tag{2.12}$$

On the other hand, if the time series used to train the encoder is gener-ated using molecular dynamics governed by potential energy function $U = U(z)$, then an additional loss term in $\dot{x}$ of the form

$$\mathcal{L}_U(\mu, K) = \left\| K(x(t)) - \nabla_z x(t) \left[-\int_0^t \nabla_z U(z(t'))dt' + \dot{z}(0) \right]_{z(t') = \mu(x(t'))} \right\|^2 \tag{2.13}$$

needs to be incorporated to optimize the autoencoder and its associated propagator K.

2.5 TOPOLOGICAL DYNAMICS ON LATENT MANIFOLDS: METAMODELS WITHOUT EQUATIONS

The sheer complexity of the molecular dynamics arising in soft and biological matter, especially in *in vivo* settings where $N \sim 10^6 - 10^7$ including solvent molecules, is unlikely to ultimately allow for the type of reductive approach that standard autoencoders usually provide. A case in point is the discovery of the physical underpinnings of protein folding assisted by an *in vivo* context that enhances the expediency of the process (Figure 2.5) [19]. The space is often anisotropic, the system itself is highly

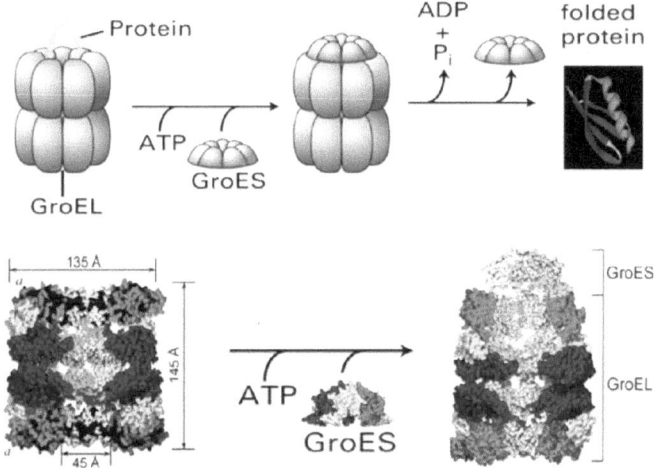

FIGURE 2.5 *In vivo* setting assisting the folding of a protein chain. A molecular cage known as chaperone GroEL assists the folding process, enhacing its expediency well above the level of efficiency that may be achieved *in vitro*, that is, in the test tube (cf. Chapter 3). The cage consists of a dimer of two annular molecular assemblages, each consisting of a complex of seven identical proteins. Each protein is made of three regions known as apical, intermediate and equatorial domain, denoted respectively as "a, i, e" in the figure. These domains undergo a certain amount of conformational rearrangement upon binding to the cell-fuel molecule ATP (adenine triphosphate). This rearrangement, in turn, enables the cage dimer to incorporate a molecular lid, known as GroES. Once the lid is on, the protein inside the cage is subject to a number of iterative annealing steps through interactions with the proteins lining the interior of the cage. This annealing process enables the protein to avoid getting kinetically trapped in misfolded states, as shown in Chapter 3. Upon release of ATP, the lid becomes detached and the protein exits the system regardless of whether it has satisfactorily completed the folding process or not. If the latter is the case, additional catalytic cycles engaging the same or other folding assistants may be necessary.

heterogeneous, and its components are too diverse, with potentials or force fields that cannot fully account for the complexities and many-body effects enshrined in the time series. It is doubtful that the latent compact manifold spanning soft-mode coordinates (backbone torsional dihedrals in the case of the folding protein; see Figure 2.4) will be amenable to the type of model discovery that is usually cast in terms of sparse differential equations. For example, generating a minimal set of 2M (M~100) coupled differential equations that govern the backbone torsional dynamics underlying the protein folding process with implicit treatment of the environment is out of reach given current capabilities in deep learning.

Other many-body systems share similar problems, as their wanton complexity is off limits for state-of-the-art autoencoders seeking to identify parsimonious models with differential equations. Yet, as we shall now show, a generic topological understanding of the latent dynamics may yield a way of learning dynamic data that enables suitable propagation of the time series into the future, endowing the autoencoder with predictive value. Thus, topological methods will be readily incorporated in AI-based model discovery for systems with multiscale complexity.

The approach entails a simplification based on the topological dynamics in the latent manifold, that is, on the dynamics modulo the basins of attraction of the generic singular points of the map $K : \Omega \rightarrow T\Omega$, where $T\Omega$ denotes the tangent bundle of Ω and Ω itself is assumed to be C^1-differentiable and compact [20], as it is the case in the previously discussed example.

To make further progress with the argument, we first prove the following result:

Theorem 2.2

Under the assumptions of Section 2.3, the autoencoded map K yields no closed orbits in Ω.

By reductio ad absurdum, let us assume $x(0) = x(T)$ for $T > 0$. Then, we get

$$0 = \int_0^T \tilde{U}(x(t))dt = \oint d\left[\int_0^t \tilde{U}(x(t'))dt'\right] == \int_0^T \left[\nabla_x \int_0^t \tilde{U}(x(t'))dt'\right]\dot{x}(t)dt$$

$$= \int_0^T \left[\int_0^t \nabla_x \tilde{U}(x(t'))dt'\right]\dot{x}(t)dt = -\int_0^T ||\tilde{v}||^2 dt \leq 0$$

$$(2.14)$$

Equation 2.13 implies that $\tilde{v} = 0$, which is absurd since $x(0)$ was not assumed to be a steady state but a point in a closed orbit. QED.

Given that Ω is a compact manifold, this result has far-reaching consequences [20, 21]:

1. The latent dynamics have no attractors made up of recurrent orbits, and

2. since there is no circulation around them, all singular points of the latent flow are hyperbolic, hence generic, since the real part of the eigenvalues of the Jacobian at the singular points cannot be zero.

This implies the following result:

Corollary 2.1

The latent dynamics governed by Equations (2.11) are of the Morse–Smale type [21], hence structurally stable, that is, qualitatively (topologically) invariant under small perturbations.

To rigorously define structural stability, we first note that the space $\mathfrak{H}(\Omega)$ of smooth maps $\Omega \to T\Omega$ is endowed with a natural metric inherited from the supremum norm given by

$$
\begin{aligned}
\|H\|_{\text{sup}} &= Sup_{x \in \Omega, j=1,\ldots,\dim\Omega} \left[\left| H(x) \right|, \left| \frac{\partial}{\partial x_j} H(x) \right| \right] \\
&= \max_{x \in \Omega, j=1,\ldots,\dim\Omega} \left[\left| H(x) \right|, \left| \frac{\partial}{\partial x_j} H(x) \right| \right]
\end{aligned}
\tag{2.15}
$$

for $H \in \mathfrak{H}(\Omega)$. Then, to state that the latent flow $K(x)$ is structurally stable means that for any given Δ-neighborhood of K, $\mathfrak{B}_\Delta(K) \subset \mathfrak{H}(\Omega)$, there exists a value $\varepsilon = \varepsilon(\Delta)$, such that for any $G \in \mathfrak{B}_\Delta(K)$ there exists an ε-homeomorphism $h_\varepsilon : \Omega \to \Omega$ satisfying $\max_{x \in \Omega} |x - h_\varepsilon(x)| < \varepsilon$ that transforms trajectories of K onto trajectories of G [21].

Given the qualitative invariance of the latent flow under small perturbations, the following observation is key to justify the leverage of AI to construct dynamic models based on time series: *The structural stability of the latent dynamics is essential to enable model discovery in view of the fact that the exact parameters determining the potential energy U of the many-body system are not known precisely.*

Given the characterization of the latent flow given by Theorem 2.2 and its corollary 2.1, we may encode the flow in a simplified manner, as we now build a metamodel. To that effect, we first define the equivalence relation "~" for any pair $x, y \in \Omega : x{\sim}y \Leftrightarrow \omega(x) = \omega(y)$, where $\omega(x)$ denotes destiny (omega) state given by the limit at $t \to \infty$ of the trajectory initiated at x [22]. Since the singular points are the steady states of the system, the latent flow may be encoded by the equivalence classes identified as the basins of attraction of the singular points. As points in Ω are regarded "modulo basins," we have in effect defined the quotient space $\Omega/{\sim}$ as the set of basins of attraction and separatrices (basins of lower dimension) of critical generic points that partition the manifold. The quotient space is relatively simple to encode since the singular points of the latent flow are isolated and finite in number. To demonstrate this proposition, we note that otherwise, if they were infinite in number, they would have an accumulation point since the latent manifold is compact, and that cannot happen because all singular points are generic.

Let us then denote by $\Gamma : \Omega/{\sim} \to \Omega/{\sim}$ the coarse-grained flow that determines the interbasin transitions, while $\pi : \Omega \to \Omega/{\sim}$ denotes the canonical projection that associates a point in the latent manifold with its equivalence class. The flow must be such that the diagram in Figure 2.6 becomes

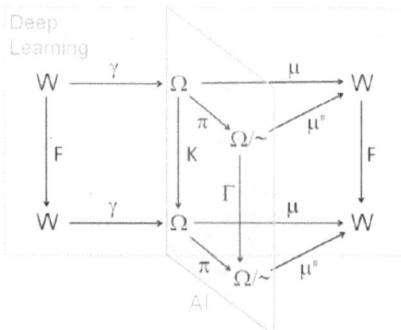

FIGURE 2.6 Scheme of a metamodel consisting of two coupled variational autoencoders (γ, μ), $(\pi, \mu^{\#})$ required to generate the discrete flow Γ that propagates the topological dynamics in the latent quotient manifold $\Omega/{\sim}$. The parameter optimization of both autoencoders ensures the full commutativity of the diagram. The discovery of the topological metamodel $\Gamma : \Omega/{\sim} \to \Omega/{\sim}$ hinges on a CNN-based construction of the modulo-basin projection $\pi : \Omega \to \Omega/{\sim}$ of the latent dynamics. This scheme introduces a level of coarse-graining that is more drastic than that adopted by conventional autoencoders for model discovery of time series data.

commutative, and, specifically, the following flow-compatibility relations must hold:

$$\mu^{\#} \circ \pi = \mu \qquad (2.16)$$

$$\Gamma \circ \pi = \pi \circ K \qquad (2.17)$$

$$\mu^{\#} \circ \Gamma = F \circ \mu^{\#} \qquad (2.18)$$

Thus, to parametrize Γ, we need to introduce a second autoencoder with variational parameter optimization determined by the loss function

$$\mathfrak{L}_{\sim}\left(\mu^{\#}, \Gamma\right) = \mathfrak{L}_{\mu}\left(\mu^{\#}\right) + \mathfrak{L}_{\sim}\left(\Gamma\right) + \mathfrak{L}_{\sim}\left(\mu^{\#}\right), \qquad (2.19)$$

where:

$$\mathfrak{L}_{\mu}\left(\mu^{\#}\right) = Q^{-1} \sum_{x \in \mathfrak{F}} \left\|\left(\mu^{\#} \circ \pi - \mu\right)x\right\|^2, \qquad (2.20)$$

$$\mathfrak{L}_{\sim}\left(\Gamma\right) = Q^{-1} \sum_{x \in \mathfrak{F}} \left\|\left(\Gamma \circ \pi - \pi \circ K\right)x\right\|^2, \qquad (2.21)$$

$$\mathfrak{L}_{\sim}\left(\mu^{\#}\right) = Q^{-1} \sum_{x \in \mathfrak{F}} \left\|\left(\mu^{\#} \circ \Gamma - F \circ \mu^{\#}\right)\pi x\right\|^2, \qquad (2.22)$$

where the training set is denoted $\mathfrak{F} \subset \Omega$, with $Q = |\mathfrak{F}|$, and $\mu^{\#}$ denotes the decoder for the quotient space.

The commutativity of the diagram in Figure 2.6 implies that the ultimate simplicity in a model governing ultra-complex many-body problems, such as identifying *in vivo* protein folding trajectories (cf. Chapter 3), may be achieved by projecting the latent dynamics onto the quotient manifold $\Omega/\sim$.

2.6 UNRAVELING TOPOLOGICAL QUOTIENT SPACES FOR DYNAMICAL SYSTEMS METAMODELING

The sparse latent dynamics obtained by leveraging autoencoders that serve as model discoverers has become the subject of intense research in applied mathematics. Such methods are less suited to unravel underlying laws in dynamical systems that represent biological or soft matter, where the number of internal degrees of freedom is astronomical. This book proposes to couple two commutative – hence compatible – autoencoders as described in Figure 2.6, yielding a factorization of the latent dynamics through the quotient space $\Omega/\sim$. The commutativity of the whole diagram displayed in Figure 2.6 ensures the compatibility of the different levels of coarse-graining of the dynamics that constitute the metamodel. In turn, the diagram commutativity (Equations 2.16–2.18) is subsumed into the variational functional of the autoencoder (Equations 2.19–2.22) so that optimization of the underlying neural networks is equated with – or rather becomes as close as possible to – diagram commutativity. Identifying the quotient space $\Omega/\sim$ under the "modulo basin" equivalence relation defined in the previous section is akin to a pattern recognition process, where time series datapoints are plotted onto 2D cross sections of the latent manifold Ω. The task may be entrusted to a CNN (Chapter 1) and becomes enormously simplified as Corollary 2.1, jointly with Theorem 2.2, guarantee a finite number of basins of attraction for isolated singular points which are all generic, that is, topologically equivalent under small perturbations of the latent vector field.

For example, in the case of the folding protein, the 2D cross section is the (Φ,Ψ)-torus (Figure 2.4(a)), and a typical time series (see Chapter 3 for details) for latent dynamics governed by Equation 2.11 is given in Figure 2.7 [9, 12]. The CNN "discovers" the topological metamodel which can be represented as a graph with vertices corresponding to the basins of attraction of the generic minima and edges connecting basins of attraction (Figure 2.7) in a manner specified by the autoencoder-generated information fed onto the CNN. More specifically, the graph is generated in accord with the inferred topography of the 2-torus cross section of the potential energy map $\tilde{U}:\Omega \to \mathbb{R}$, in turn generated by the autoencoder.

To generate the entire quotient space within a graph representation, we need to integrate the cross sections corresponding to the 2D-projections of $\Omega/\sim$. As an illustration, let us consider a GG dipeptide (M = 2, G = glycine). The latent manifold is a Cartesian product of two tori (Figure 2.8), one for each pair of Φ, Ψ dihedral coordinates specifying the backbone conformation of the respective residue [12]. The topological representation of the vector field steering the backbone torsional dynamics of a generic residue in the protein chain is given in Figure 2.7. Opposite sides of the square are identified as per the ±180° identification of Φ, Ψ dihedrals determining the local torsional state of the backbone. The four colored

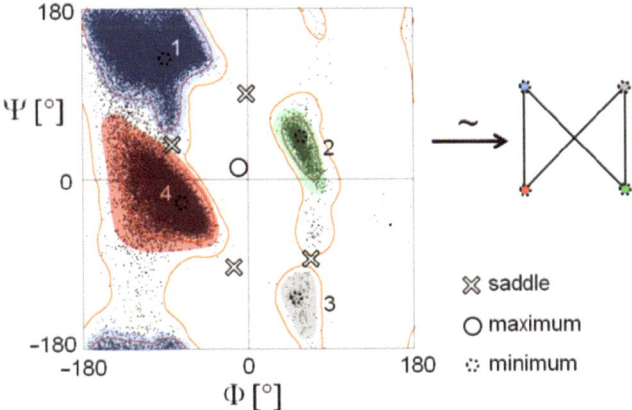

FIGURE 2.7 Toroidal (Φ, Ψ) cross section of the time series data for an individual unit along a folding protein chain. The associated cross section of the dynamical system representing the protein folding process is discovered through pattern recognition leveraging a CNN, and its modulo-basin representation in the cross section of the latent quotient manifold is given as a graph. The vertices in the graph are critical points corresponding to omega sets for all points in their respective basins of attraction, while the edges indicate allowed interbasin transitions that determine the topological dynamics.

sectors morph topologically into the allowed valleys in potential energy. The organization of the basins of critical points is compatible with the underlying 2-torus and topologically represented by a graph (Figure 2.7) for each residue. The bottom panel in Figure 2.8 represents the quotient space for the dipeptide chain [12]. For each residue, the quotient space cross section is represented by a graph with vertices indicating 2-dimensional basins, and an edge linking two vertices indicates that the respective basins are connected through a line of steepest descent crossing a saddle point and orthogonal to the separatrix at the saddle. For a protein chain consisting of two consecutive residues, denoted 1 and 2, the quotient space becomes the tensor product of the two graphs where vertices now denote basin pairs (B_1,B_2), and where basin pair (B_1,B_2) is connected with (B'_1, B'_2) if and only if B_1 is connected with B'_1 or $B_1=B'_1$, and B_2 is connected with B'_2 or $B_2=B'_2$. Thus, the quotient space for two residues consists of $4 \times 4 = 16$ vertices, where each vertex connects via one edge with 8 other vertices and connects via two adjacent edges to the 7 remaining vertices. Given the symmetry of the problem, to prove the assertion it suffices to note that the vertex denoting basin pair (1,1) is directly

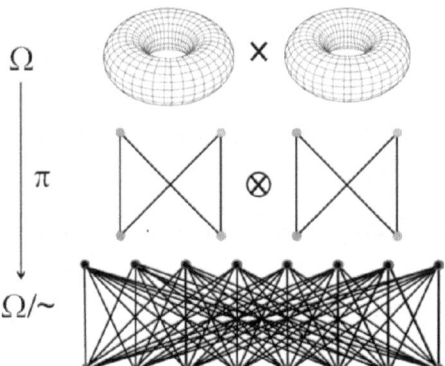

FIGURE 2.8   Reconstruction of the latent quotient manifold. The iterative process is represented by progressively incorporating adjacent-residue cross sections of the latent manifold as Cartesian products and, in parallel, reconstructing the quotient manifold as tensorial product of the individual graphs representing the cross sections of the quotient manifold (Figure 2.7). Within this representational framework, the topological dynamics becomes a walk in the quotient manifold graph.

connected to 8 other vertices denoting pairs (2,1), (4,1), (1,2), (1,4), (2,2), (2,4), (4,2), (4,4) while vertex (1,1) connects to all the remaining seven vertices-pairs containing basin 3 via two adjacent edges.

Thus, the projection on quotient space $\Omega/\sim$ of a latent MD trajectory on the 4-torus Ω becomes a walk on the tensorial product graph at the bottom of Figure 2.8 [12].

2.7 LEARNING TO ENCODE AND PROPAGATE TOPOLOGICAL DYNAMICS: METAMODEL FOR UBIQUITIN FOLDING IN THE CELL

The coupling of two autoencoders where all flows commute as described in Figure 2.6 becomes essential to discover topological metamodels governing the ultra-complex dynamics encountered in biological matter. Such systems are unlikely to yield sparse differential equations governing the latent dynamics in a compact manifold but that is only an educated guess. To train both autoencoders in parallel, it is necessary to project the latent dynamics generated by Equation (2.11) onto the modulo-basin topological dynamics defined on the quotient space and use the encoded information to optimize the autoencoders according to the variational functionals (2.3–2.7) and (2.19–2.22) defined in the previous sections.

As an illustration, Figure 2.9(a) describes a latent state of the protein chain with chemical composition (amino acid sequence) corresponding to the human *ubiquitin* [9]. The state is represented as a point in the latent compact manifold $\Omega = \Pi_{n=1}^{M}\Omega_n$ ($M = 76$), where Ω_n is the 2-torus

	1	2	3	4	5	6	7	8	9	10
Amino acid	MET-M	GLN-Q	ILE-I	PHE-F	VAL-V	LYS-K	THR-T	LEU-L	THR-T	GLY-G
Phi-angle	56.50	-92.00	-129.00	-111.00	-115.00	-89.00	-95.00	-73.00	-94.00	88.00
Psi-angle	34.92	133.00	160.00	134.00	106.00	121.00	170.00	-11.00	3.00	14.00
Omega-angle	180.00	180.00	180.00	180.00	180.00	180.00	180.00	180.00	180.00	180.00
	11	12	13	14	15	16	17	18	19	20
Amino acid	LYS-K	THR-T	ILE-I	THR-T	LEU-L	GLU-E	VAL-V	GLU-E	PRO-P	SER-S
Phi-angle	-102.00	-112.00	-108.00	-96.00	-128.00	-107.00	-137.00	-116.00	-55.00	-83.00
Psi-angle	131.00	128.00	136.00	138.00	150.00	114.00	168.00	142.00	-16.00	-45.00
Omega-angle	180.00	180.00	180.00	180.00	180.00	180.00	180.00	180.00	180.00	180.00
	21	22	23	24	25	26	27	28	29	30
Amino acid	ASP-D	THR-T	ILE-I	GLU-E	ASN-N	VAL-V	LYS-K	ALA-A	LYS-K	ILE-I
Phi-angle	-74.00	-81.00	-64.00	-57.00	-71.00	-62.00	-61.00	-68.00	-64.00	-64.00
Psi-angle	143.00	161.00	-37.00	-39.00	-37.00	-43.00	-34.00	-37.00	-42.00	-39.00
Omega-angle	180.00	180.00	180.00	180.00	180.00	180.00	180.00	180.00	180.00	180.00
	31	32	33	34	35	36	37	38	39	40
Amino acid	GLN-Q	ASP-D	LYS-K	GLU-E	GLY-G	ILE-I	PRO-P	PRO-P	ASP-D	GLN-Q
Phi-angle	-64.00	-58.00	-94.00	-118.00	84.00	-79.00	-58.00	-56.00	-64.00	-92.00
Psi-angle	-43.00	-38.00	-27.00	-11.00	5.00	119.00	136.00	-35.00	-18.00	-10.00
Omega-angle	180.00	180.00	180.00	180.00	180.00	180.00	180.00	180.00	180.00	180.00
	41	42	43	44	45	46	47	48	49	50
Amino acid	GLN-Q	ARG-R	LEU-L	ILE-I	PHE-F	ALA-A	GLY-G	LYS-K	GLN-Q	LEU-L
Phi-angle	-85.00	-128.00	-95.00	-122.00	-143.00	51.00	72.00	-112.00	-84.00	-75.00
Psi-angle	130.00	107.00	131.00	130.00	128.00	43.69	-21.00	140.00	122.00	134.00
Amino acid	180.00	180.00	180.00	180.00	180.00	180.00	180.00	180.00	180.00	180.00
	51	52	53	54	55	56	57	58	59	60
Amino acid	GLU-E	ASP-D	GLY-G	ARG-R	THR-T	LEU-L	SER-S	ASP-D	TYR-Y	ASN-N
Phi-angle	-95.00	-45.00	-75.00	-82.00	-103.00	-61.00	-66.00	-58.00	-97.00	63.00
Psi-angle	133.00	-40.00	-20.00	164.00	165.00	-33.00	-31.00	-32.00	-1.00	34.00
Omega-angle	180.00	180.00	180.00	180.00	180.00	180.00	180.00	180.00	180.00	180.00
	61	62	63	64	65	66	67	68	69	70
Amino acid	ILE-I	GLN-Q	LYS-K	GLU-E	SER-S	THR-T	LEU-L	HIS-H	LEU-L	VAL-V
Phi-angle	-77.00	-93.00	-53.00	73.00	-74.00	-117.00	-97.00	-105.00	-105.00	-112.00
Psi-angle	112.00	169.00	140.00	18.00	157.00	122.00	151.00	129.00	117.00	133.00
Omega-angle	180.00	180.00	180.00	180.00	180.00	180.00	180.00	180.00	180.00	180.00
	71	72	73	74	75	76				
Amino acid	LEU-L	ARG-R	LEU-L	ARG-R	GLY-G	GLY-G				
Phi-angle	-91.00	-122.00	-111.00	-111.00	-56.99	-64.99				
Psi-angle	138.00	95.00	120.00	120.00	-46.01	-59.81				
Omega-angle	180.00	180.00	180.00	180.00	180.00	180.00				

(a)

FIGURE 2.9 Latent manifold and latent quotient manifold for a folding protein. **(a)** Dihedral torsional representation of a point in the latent manifold for the folding of ubiquitin (M=76, dimΩ=152).

(Continued)

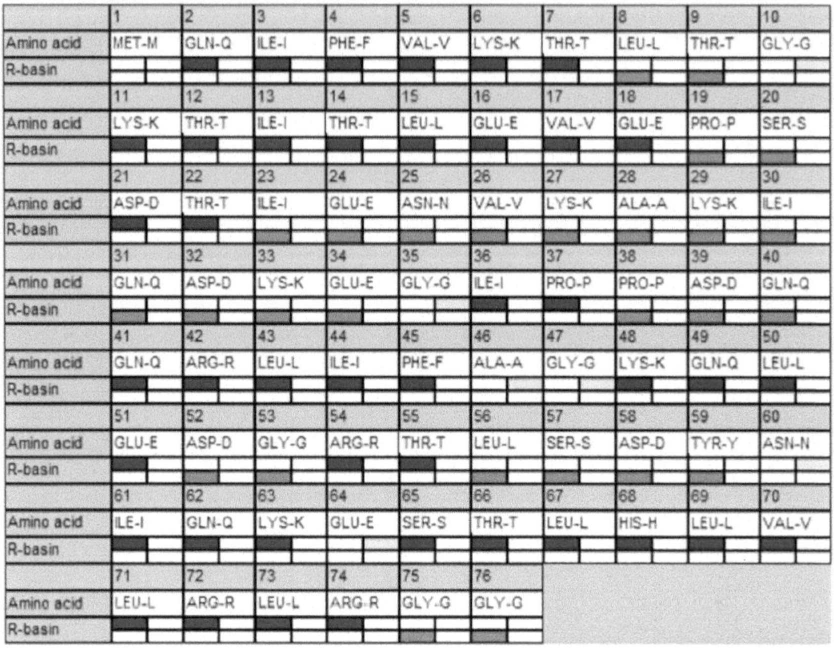

	1	2	3	4	5	6	7	8	9	10
Amino acid	MET-M	GLN-Q	ILE-I	PHE-F	VAL-V	LYS-K	THR-T	LEU-L	THR-T	GLY-G
R-basin										
	11	12	13	14	15	16	17	18	19	20
Amino acid	LYS-K	THR-T	ILE-I	THR-T	LEU-L	GLU-E	VAL-V	GLU-E	PRO-P	SER-S
R-basin										
	21	22	23	24	25	26	27	28	29	30
Amino acid	ASP-D	THR-T	ILE-I	GLU-E	ASN-N	VAL-V	LYS-K	ALA-A	LYS-K	ILE-I
R-basin										
	31	32	33	34	35	36	37	38	39	40
Amino acid	GLN-Q	ASP-D	LYS-K	GLU-E	GLY-G	ILE-I	PRO-P	PRO-P	ASP-D	GLN-Q
R-basin										
	41	42	43	44	45	46	47	48	49	50
Amino acid	GLN-Q	ARG-R	LEU-L	ILE-I	PHE-F	ALA-A	GLY-G	LYS-K	GLN-Q	LEU-L
R-basin										
	51	52	53	54	55	56	57	58	59	60
Amino acid	GLU-E	ASP-D	GLY-G	ARG-R	THR-T	LEU-L	SER-S	ASP-D	TYR-Y	ASN-N
R-basin										
	61	62	63	64	65	66	67	68	69	70
Amino acid	ILE-I	GLN-Q	LYS-K	GLU-E	SER-S	THR-T	LEU-L	HIS-H	LEU-L	VAL-V
R-basin										
	71	72	73	74	75	76				
Amino acid	LEU-L	ARG-R	LEU-L	ARG-R	GLY-G	GLY-G				
R-basin										

(b)

FIGURE 2.9 *(Continued)* **(b)** Equivalence class in the latent quotient manifold containing the point given in (a). The four basins of attraction are represented by quadrants in a square.

spanned by the two backbone torsional dihedral coordinates of the nth residue along the chain. Using the modulo-basin encoding $\pi : \Omega \to \Omega/\sim$ defined in Figure 2.8, the state given in Figure 2.9(a) becomes part of the equivalence class in $\Omega/\sim$ defined by the modulo-basin topology given in Figure 2.9(b). This can be verified residue by residue using Figure 2.7. Once the two coupled autoencoders are trained in parallel, the system learns to propagate dynamics in the latent quotient space $\Omega/\sim$, as described in Figure 2.10. The training process in this case involved a laborious 0.5 ms-MD simulation of ubiquitin folding assisted by the chaperone GroEL (Figure 2.5), as detailed in the Chapter 3. The decoding $\mu^{\#} : \Omega/\sim \to W$ of the modulo-basin topology generated by the learned flow Γ at 7 ms yields a protein structure (Figure 2.10) which is very close (RMSD = 1.15 Å) to the native fold reported in the entry 1UBI of the protein data bank (PDB). This experimental validation attests to the power of leveraging topological dynamics for AI-enabled model discovery of biological matter.

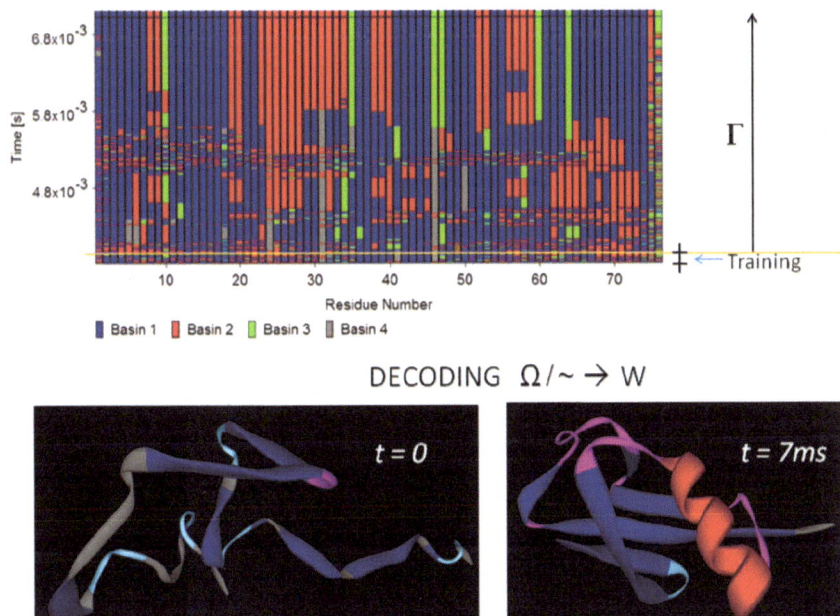

FIGURE 2.10   Latent folding dynamics generated by two coupled autoencoders (Figure 2.6) yielding an *in vivo*-assisted pathway that expeditiously yields the native fold of ubiquitin (PDB.1UBI). The autoencoders are trained by time series obtained by running MD trajectories spanning 0.5 ms of a molecular *in vivo* setting (chaperone, Figure 2.5) that steers the folding chain. Each horizontal line in the plot represents a modulo-basin conformation of the entire chain resolved at time intervals of 50 μs. Hence, the plot describes the AI-empowered propagation of the folding dynamics projected on the latent quotient space. The plot constitutes a topological dynamics metamodel of the folding of ubiquitin.

2.8 METAMODELS FOR HIERARCHICAL DYNAMICS DISCOVERED THROUGH AUTOENCODER BATTERIES

Kurzweil and others have successfully argued that a hierarchical structure of reality is necessary for proper encoding and processing in a suitable AI-based inferential framework [10, 23]. To cast the discussion in the broadest terms, we shall refer to hierarchical as the attribute of a system endowed with nested complexities at multiple scales arising at different levels of description and providing different levels of coarse-graining. This notion was delineated in Chapter 1, where a number of illustrations reveal the dynamic entrainment of fast-relaxing modes by slower modes spanning a latent manifold. Thus, just like in the adiabatic approximation

[14], fast motions are averaged out and hence treated implicitly in a coarse-grained version of the dynamics focusing on longer timescales, with an autoencoder providing the inferential framework to discover the latent coordinate system.

Indeed, an escalation in the level of coarse-graining, from the sub-atomic to the atomic, molecular, subcellular and beyond, has always suggested that the hierarchical structure of reality may span over several layers, with nested complexities where information at a peripheral layer is incorporated implicitly at a core layer. In topological terms, we envision a whole sequence of quotient manifolds, as we lump up states at different levels of description within progressively coarser equivalence classes, with dynamic compatibility of the various descriptions imposed by commutative flow diagrams (cf. Figure 2.11(a)). Thus, in the broadest sense, an encoding of a hierarchical dynamical system within an AI-based inferential framework may require a battery of tandem autoencoders, whose dynamic compatibility in an optimized parametrization is ensured by the

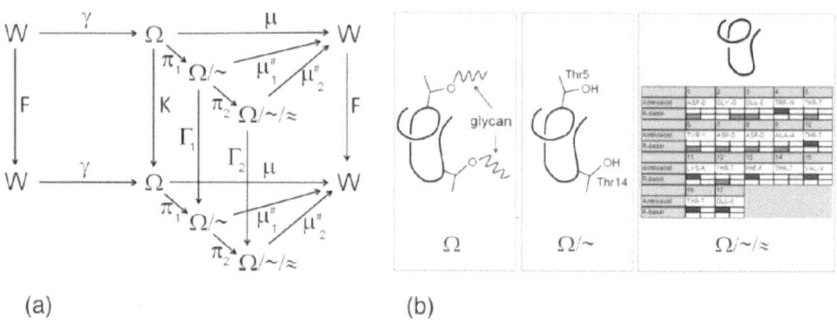

(a)   (b)

FIGURE 2.11 Metamodels for hierarchical dynamical systems. Autoencoders in tandem implemented to forecast dynamics that allow for a hierarchical description defined by two equivalence relations "~" and "≈." (a) Two autoencoders {π_1, $\mu^\#_1$, Γ_1} and {π_2, $\mu^\#_2$, Γ_2} operating sequentially as required for topological encoding of the dynamics at the coarsest level in the latent quotient manifold $\Omega/\sim/\approx$. The F-compatibility of the latent flows K, Γ_1, Γ_2 in $\Omega, \Omega/\sim$, and $\Omega/\sim/\approx$, respectively, requires the full commutativity of the diagram. This commutativity is achieved by variational parameter optimization for the three autoencoders {γ, μ, K}, {π_1, $\mu^\#_1$, Γ_1}, and {π_2, $\mu^\#_2$, Γ_2}. (b) Hierarchical description of the state of a protein chain subject to post-translation chemical modification in the form of glycosylation. Two glycosylated states are regarded as equivalent if and only if the underlying protein chain conformations are identical as the attached glycans are disregarded. The latent manifolds $\Omega, \Omega/\sim$ and $\Omega/\sim/\approx$ indicate a progression in coarse-graining with the quotient $\Omega/\sim/\approx$ containing the "modulo basin" topologies.

commutativity of the diagram that combines the different levels of flow encoding. Figure 2.11(a) shows one such diagram reflecting the dynamic interplay of three autoencoders, one to generate the latent manifold Ω of intrinsic coordinates, one for a first-level quotient space $\Omega/\sim$, and one for a second-level (coarser) quotient space $\Omega/\sim/\approx$. The autoencoders are variationally optimized so that the flow diagram becomes commutative, which in turn reflects the compatibility of the different levels of dynamic encoding within the latent 3-level hierarchy.

Let us illustrate the hierarchical metamodel discovery by critically revisiting the protein folding dynamics in its *in vivo* setting discussed in the previous section. Like almost everybody else, in the previous discussion we overlooked the fact that the protein undergoes significant chemical post-translational modification that alters its chemical composition and thereby its folding dynamics [9, 16]. In fact, protein chains undergo extensive glycosylation at specific sites for specific residues, predominantly serine (Ser, S), threonine (Thr, T) and tyrosine (Tyr, Y) [24]. The glycosylation process entails covalent attachment of carbohydrate-based oligomers called glycans at side-chain hydroxyl (-OH) groups of S, T or Y. The folding process is critically affected by the high polarity and high entropy content due to dihedral torsional freedom of the glycans [24]. By introducing the equivalence relation "$\sim$," we identify glycosylated chain conformations as equivalent if the underlying protein chain conformations are identical when the attached glycans are removed (Figure 2.11(b)). Thus, the folding dynamics of the glycosylated chain determined by flow F may be predicted/propagated for future times by the autoencoder $\{\pi, \mu^{\#},$ $\Gamma\}$ that is variationally optimized to make the diagram in Figure 2.6 commutative while treating implicitly the influence of glycosylation on folding as per the quotient $\Omega/\sim$. The Γ-dynamics on $\Omega/\sim$ is further entrained by the "modulo basin" topological dynamics at $\Omega/\sim/\approx$, where the equivalence relation "$\approx$" identifies protein chain conformations that share the same assignment of dihedral torsional basin to each residue (Figure 2.11(b)).

Adopting a dummy subindex to denote the respective autoencoders, the hierarchical reduction is operationally materialized through tandem combination of autoencoders $\{\pi_1, \mu^{\#}_1, \Gamma_1\}$ and $\{\pi_2, \mu^{\#}_2, \Gamma_2\}$ that hierarchically account for the dynamics in quotient manifolds $\Omega/\sim$ and $\Omega/\sim/\approx$, respectively. The variational parameter optimization of the two autoencoders ensures the commutativity of the diagram in Figure 2.11(a). In this way, the hierarchical dynamics may be predicted and propagated by leveraging the autoencoders to operate in tandem and decoding the generated

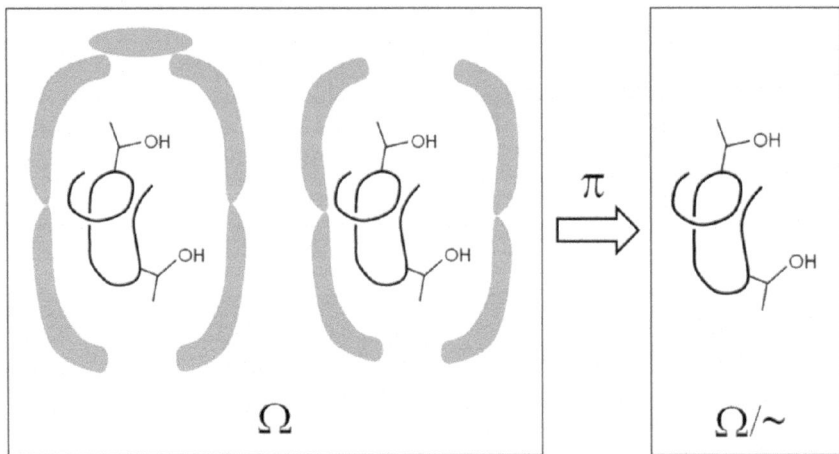

FIGURE 2.12 Hierarchical description of the state of a folding protein chain assisted by an *in vivo* context consisting of a protein assemblage (gray regions) which interacts with the folding chain. The autoencoder associated with the equivalence relation "~" learns to incorporate implicitly the influence of the *in vivo* context as it propagates the folding dynamics encode the latent manifold $\Omega/\sim$.

topological dynamics that underlies the process at the coarsest level in the latent quotient manifold $\Omega/\sim/\approx$. There is a significant computational and conceptual advantage in discovering the decodable coarsest (topological) dynamics, since it represents the utmost modeling simplicity. This constitutes the metamodel for hierarchical dynamics.

Similarly, the participation of an *in vivo* context assisting the protein folding process may be incorporated implicitly by adopting hierarchically coupled autoencoders. Thus, two states of the chain confined within their respective *in vivo* protein assemblages are regarded as equivalent under the relation "~" if the conformation of the chain itself is identical as the conformational states of the caging assemblage are disregarded (Figure 2.12). Thus, both chain states are projected onto the same class modulo "~." The autoencoder for the equivalence "~" is trained to fold the chain *in vivo* by treating the *in vivo* context implicitly, with the training set consisting of time series data generated by molecular dynamics runs on the folding chain jointly with its confining *in vivo* assemblage. None of these runs covers physically realistic timescales but taken together they instruct the autoencoder on how to implicitly encode the *in vivo* context.

The subsequent chapters illustrate the implementation of tandem autoencoder technology to treat implicitly the hierarchical structure of the molecular dynamics of biological and biomedical processes. The relationship between hierarchical structure and tandem autoencoders of nested coarse-graining descriptions is of universal applicability and may be extended to any number of equivalence relations, leading to progressively simplified metamodels.

REFERENCES

1. Russell S, Norvig P (2020) *Artificial intelligence: A modern approach*. Pearson, London, UK.
2. Kelleher JD (2019) *Deep learning*. The MIT Press, Cambridge, MA.
3. Chollet F (2019) *Deep learning with python*. Manning Press, Shelter Island, New York.
4. Brunton SL, Kutz NJ (2019) *Data-driven science and engineering: Machine learning, dynamical systems and control*. Cambridge University Press, UK.
5. Tiumentsev Y, Egorchev M (2019) *Neural network modeling and identification of dynamical systems*. Academic Press, San Diego, USA.
6. Champion K, Lusch B, Kutz JN, Brunton SL (2019) Data-Driven Discovery of Coordinates and Governing Equations. *Proc Natl Acad Sci USA* 116: 22445–22451.
7. Schmidt M, Lipson H (2009) Distilling Free-Form Natural Laws from Experimental Data. *Science* 324:81–85.
8. Brunton SL, Proctor JL, Kutz NJ (2016) Discovering Governing Equations from Data by Sparse Identification of Nonlinear Dynamical Systems. *Proc Natl Acad Sci USA* 113:3932–3937.
9. Fernández A (2021) *Artificial intelligence platform for molecular targeted therapy: A translational approach*. World Scientific Publishing, Singapore.
10. Kurzweil R (2006) *The singularity is near: When humans transcend biology*. Penguin, New York.
11. de Vries J (2014) *Topological dynamical systems: An introduction to the dynamics of continuous mappings*. De Gruyter, Berlin.
12. Fernández A (2020) Deep Learning Unravels a Dynamic Hierarchy While Empowering Molecular Dynamics Simulations. *Ann Physik (Berlin)* 532:1900526.
13. Duraisamy K, Iaccarino G, Xiao H (2018), Turbulence modeling in the age of data. *Annu Rev Fluid Mech* 51:357–377.
14. Fernández A (1985) Center-Manifold Extension of the Adiabatic-Elimination Method. *Phys Rev A* 32:3070–3076.
15. Thommen M, Holtkamp W, Rodnina MV (2017) Co-Translational Protein Folding: Progress and Methods. *Curr Opin Struct Biol* 42:83–89.
16. Sorokina I, Mushegian A (2018) Modeling Protein Folding In Vivo. *Biology Direct* 13:13.

17. Thurston WP (1997) *Three-Dimensional Geometry and Topology*. Princeton University Press, Princeton, NJ.

18. Brooks B, Karplus M (1983) Harmonic Dynamics of Proteins: Normal Modes and Fluctuations in Bovine Pancreatic Trypsin Inhibitor. *Proc Natl Acad Sci U S A* 80:6571–6575.

19. Thirumalai D, Lorimer GH, Hyeon C (2020) Iterative Annealing Mechanism Explains the Functions of the GroEL and RNA Chaperones. *Protein Sci* 29:360–377.

20. Nemytskii VV, Stepanov V (2016) *Qualitative theory of differential equations*. Princeton University Press, Princeton, NJ.

21. Arnold VI (1974) *Mathematical methods of classical mechanics*. Springer, Berlin.

22. Weisstein EW (2021) Quotient Space. From MathWorld--A Wolfram Web Resource. https://mathworld.wolfram.com/QuotientSpace.html

23. Li S, Yang Y (2021) Hierarchical Deep Learning for Data-Driven Identification of Reduced-Order Models of Nonlinear Dynamical Systems. *Nonlinear Dyn* 105:3409–3422.

24. Drickamer K, Taylor ME (2011) *Introduction to glycobiology*, 3rd Edition. Oxford University Press, UK.

II

Applications

## CHAPTER 3

Artificial Intelligence Reverse-Engineers *In Vivo* Protein Folding

The reasonable man adapts himself to the world,
the unreasonable one persists in trying to adapt
the world to himself. Therefore all progress
depends on the unreasonable man.

– George Bernard Shaw

3.1 DECONSTRUCTING *IN VIVO* PROTEIN FOLDING: TOPOLOGICAL METAMODEL CREATED BY TRANSFORMER TECHNOLOGY

Due to their wanton complexity, the majority of molecular processes that take place *in vivo* have proven unyielding to computational modeling. This situation is likely to change with the leveraging of AI technologies that have already proven capable of creating and handling hierarchical representations of complex molecular information and make accurate predictions [1]. This chapter focuses on the deployment of AI to deal with the next frontier: time-dependent molecular transformations that take place in *in vivo* settings.

Molecular dynamics (MD) is a primary tool to investigate biomolecular processes *in vitro*, such as protein folding [2, 3]. It is now believed that such efforts may have been misplaced in some instances [4–6]. For example, natural proteins actually fold and have evolved to fold in the cell, not in the test tube, with only a handful of small single-domain proteins finding their native structure *in vitro* [7–9]. With a significant proportion of proteins denaturing irreversibly *in vitro* ([6] and references therein), the Anfinsen scenario, where the protein finds its native fold *in vitro* under thermodynamic control, may require reassessment and revision, even if the Anfinsen principle asserting that structure is encoded solely by amino acid sequence remains valid [7].

Through a steering stochastic participation so far largely unidentified at the molecular level, the *in vivo* context prevents the protein from getting kinetically trapped in misfolded states and provides a cooperative context to expedite folding [4–6]. It is thus likely that the single-versus-multiple pathway controversy [10, 11] may need to be redirected to address the *in vivo* context. On its own, MD seems underequipped to reproduce such complexities at realistic catalytic turnover times. Even in the more tractable *in vitro* context, the difficulties faced by MD are apparent, as physically relevant timescales are often inaccessible, conformation space is sparsely sampled and rare events are often missed [6, 12–14]. This chapter squarely addresses these challenges, heralding the leverage of artificial intelligence (AI) technologies to empower the reverse engineering of the *in vivo* context that accounts for the expediency of the protein folding process.

The huge informational burden that needs to be carried over from one integration step to the next makes it virtually impossible for MD to recreate *in vivo* reality. To address the problem, AI becomes essential. To implement an AI-empowered MD-trained system requires that we first encode or project the protein chain dynamics onto a coarse-grained representation retaining only essential topological features of the vector field that steers the integration process [9, 12]. This mathematical approach was delineated in Chapter 2 and is now specialized to the *in vivo* protein folding problem. We simplify the flow by (a) lumping conformations within basins of attraction of potential energy minima and other singularities [9] and (b) encoding the dynamics into a modulo-basin "textual" representation that effectively averages out fast motion within the basins. Interbasin transitions in the metamodel are determined by a transformer network [15] where the receptive field for a basin transition at a residue site becomes progressively expanded within hidden layers consisting of long short-term memory (LSTM) modules parametrized by the attention range, denoted g.

Subsequently, the weighted influence of different contexts upon each basin transition is variationally optimized by contrasting transformer inferences against experimentally determined structures [16]. With a suitable propagation scheme obtained by training the transformer with big dynamic data from atomistic MD runs, this AI-empowered metamodel scheme enables coverage of realistic timescales and recreates cooperative *in vivo* reality, while atomistic level detail is recovered through AI-based decoding [16].

To enable AI to construct a metamodel of *in vivo* reality and assess its role in expediting the folding process, we incorporate the influence of specific structurally reported cellular machinery. This is accomplished by training the transformer with atomistic MD simulations of folding events within specific cellular compartments. None of the MD runs is able to cover realistic folding timescales but together they inform the transformer on the steering cooperative influence of the *in vivo* context.

For the sake of illustration, we adopt a training set consisting of MD runs that incorporate the influence of GroEL, a widely studied deca-tetrameric ATP-consuming chaperone [4], on protein folding [5]. Inside the GroEL chamber in the apo state, our transformer-enhanced atomistic MD runs reveal an intermittent alternating annealing process where noncompact states with large radius of gyration (i.e., scaling with the unfolded ensemble average) interact strongly with structurally disordered hydrophobic residues in the GroEL (GGM)-enriched C-terminus tails. These interactions are cyclically choreographed with alternating subunit participation to stochastically disrupt metastable states that would constitute kinetic traps if the folding process were to take place in bulk solvent. On the other hand, when the catalytic chamber is in the $(ATP)_7$-state, it is able to boost folding expediency via enhanced intermolecular interactions that stabilize the folding-nucleating hydrogen-bond pattern of the substrate.

Thus, the influence of the *in vivo* context is incorporated through the training of the transformer in a selected environment of molecular complexity where only sub-μs events would be accessible to standard atomistic MD computation. The transformer generates a dynamics in which two fundamental aspects of chaperone assistance become apparent: (a) coarse-grained misfolded states are levied interbasin transitions that eventually get them dismantled, thereby preventing the protein from getting kinetically trapped; and (b) the folding nucleating core gets protected stochastically through intermolecular interactions.

The intent of this chapter is to empower MD by leveraging AI to incorporate big dynamic data on cellular settings. Via feature extraction, our transformer reverse-engineers the cooperative *in vivo* context that expedites the folding process, ultimately yielding experimentally validated structures. The resulting dynamics provides mechanistic underpinnings to empirical models previously developed to fit kinetic experimental data.

3.2 EMPOWERING MOLECULAR DYNAMICS WITH TRANSFORMER TECHNOLOGY

3.2.1 Propagating the Topological Dynamics in Textual Form with a Transformer Neural Network

To reduce the informational burden in MD integration steps, we first simplify the dynamics taking into account the inherent topology of the MD-steering force field. The procedure known as quotient-space projection has been previously introduced [16–18] and is described in Chapter 2, so it will only be described cursorily for completion, emphasizing the adaptation required to implement an AI-model of the *in vivo* molecular reality. To describe the projection onto quotient space, we consider conformations described by the backbone torsional coordinates $\{\phi_n, \psi_n\}_{n=1,\dots N}$ for a protein chain of length N. Conformation space becomes a 2N-dimensional torus, that is, the product of 2N unit circles, one for each dihedral coordinate of the protein backbone [9, 12, 16–18]. To coarse-grain atomistic MD runs, we introduce an equivalence relation "~," where two conformations are regarded as equivalent if and only if the torsional states of the individual residues lie within the same basins of attraction in the local potential energy surfaces, that is, the Ramachandran plots. The Ramachandran basins are the allowed low-energy regions available to backbone torsional coordinates for each residue (Figure 3.1) [16]. The rigorous mathematical construction of the modulo-basin metamodel simplification is given in the following section.

The "modulo basin" coordinate representation constitutes a "quotient space" made up of equivalence classes. In generic terms, if x indicates the torsional state of the chain, the modulo-basin state to which x belongs is the class denoted $\bar{x}$, expressed as the Cartesian product of N Ramachandran basins designating the basin occupancies for the individual residues: $\bar{x} = \Pi_{n=1}^{N} b(x, n)$, where $b(x, n)$ is the basin of attraction occupied by residue n in conformation x. Thus, for any two conformations $x, y : x \sim y \Leftrightarrow \bar{x} = \bar{y}$ and the states in quotient space may be designated by encoding vectors. It is worth noting that the modulo-basin representation of the backbone

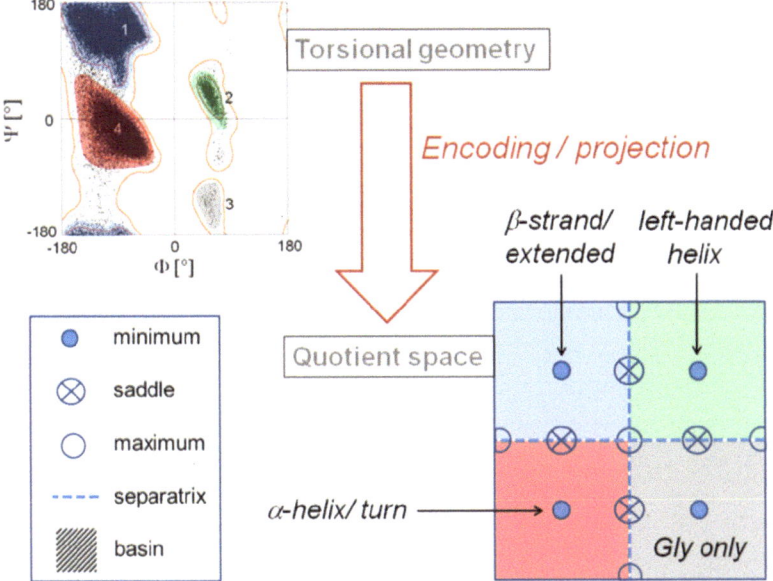

FIGURE 3.1 Modulo-basin topological representation of the backbone torsional state of a protein chain. The topological representation of the Ramachandran energy plot steering the backbone torsional dynamics of a single residue is shown in the lower panel. The allowed regions in local torsional space are displayed in the upper panel. Opposite sides of the square are identified as per the ±180° identification of ϕ, ψ dihedrals, yielding a 2-torus, or cartesian product of two circles. The four colored sectors morph topologically into the Ramachandran basins of potential energy subsuming the local conformational motifs, like helix, turn, β-strand, and so on. Except for glycine (G), the gray sector is energetically inaccessible to other residues due to steric clash with the side chain. Other accessibility restrictions apply to geometrically constrained residues like proline (P) and its adjacent residues. Thus, the local 2-torus geometry is canonically projected onto the local topology with its organization of critical points.

torsional state of the chain is a "textual" representation, where the alphabet is comprised of 4 "letters" indicating the basin assignment at each position on the chain. This text can be processed by a transformer sequence-to-sequence (Seq2Seq) architecture [15], so the sequence transduction represents a broad conformational move arising once fast motions (intrabasin transitions) are averaged out.

We may project the MD trajectory on quotient space and then adopt transformer learning to propagate the textual modulo-basin metamodel of the dynamics to make computations more agile. This text entails a considerable simplification of conformation space, while the transition timescale for each iteration is the minimal Ramachandran intrabasin equilibration time $\tau = 100$ ps, considerably larger than the femtosecond integration time step in MD. Since intrabasin events are averaged out in the modulo-basin representation and the iteration time step is five orders of magnitude longer, the propagated modulo-basin dynamics has a far better chance of capturing infrequent interbasin transitions, a significant hurdle in MD modeling of *in vitro* protein folding.

The transformer text processing needed to propagate the modulo basin dynamics uses a maximum likelihood *ansatz* to train the transformer so that the maximum likelihood weight, $f_\tau^{(t)}(n, b \rightarrow b')$, for basin transition $b \rightarrow b'$ at chain position n after time τ elapsed from time t is influenced by the amino acid identity and basin occupancy at flanking residues ($n - 2$, $n - 1$, $n + 1$, $n + 2$), with an influence field in the LSTM progressively dilated, so that flanking positions ($n - 3$, $n - 2$, $n + 2$, $n + 3$), ($n - 4$, $n - 3$, $n + 3$, $n + 4$),... , ($n - 8$, $n - 7$, $n + 7$, $n + 8$) with respective attention gaps $g = 1$, 2,..., 7, also impact the transition probability. Although both are grounded in the maximum likelihood estimation of basin transition probabilities using the same training data, this representation of the receptive field is different from a previous one that yields identical results in the *in vitro* context [17]. The previous "global" method learns to estimate transition probabilities between modulo-basin states of the whole chain, while the current approach determines basin transition probabilities at the residue level using semilocal influence fields. In this way, the number of neurons in the hidden layers is significantly reduced. This simplification is essential to reduce computational costs as the transformer learns to incorporate the complex *in vivo* reality.

Thus, the maximum likelihood weight of the basin transition becomes:

$$f_\tau^{(t)}\left(n, b \rightarrow b'\right) = \prod_{g=1}^{L} w\left(n, g, B(t)\right) f\left(n, g, b \rightarrow b'\right) \tag{3.1}$$

where $f(n, g, b \rightarrow b')$ gives the frequency of the $b \rightarrow b'$ basin transition at position n after time τ elapsed obtained by projecting 100 "short" 220 ns-MD runs onto a modulo-basin τ-discretized time series for a specific chain composition giving the amino acid identity of residues $n - g - 1$, $n - g$, n, $n + g$, $n + g + 1$ in the chain, and basin assignment for residues $n - g - 1$, $n - g$, n, $n + g$, $n + g + 1$ determined by B(t), the chain basin occupancy at time t. The different g-attention fields ($n - g - 1$, $n - g$, n, $n + g$, $n + g + 1$) bearing on the basin transition at position n are weighted in accord with attention weights $w(n, g, B(t))$, which are optimized vis-à-vis the loss function of the transformer.

A first stage in the transformer training yields transition frequencies $f(n, g, b \rightarrow b')$ following the maximum likelihood scheme. The propagation of the modulo-basin trajectory (time series) in the maximum likelihood scheme is determined by the $b \rightarrow b'$ transition probabilities for generic position n for the jump $t \rightarrow t + \tau$:

$$p_{ML}^{(t)}(n, b \rightarrow b') = \frac{f_\tau^{(t)}(n, b \rightarrow b')}{\Sigma_{b''} f_\tau^{(t)}(n, b \rightarrow b'')},$$

where b'' is any of the four basins. The destiny basin b' after a time period τ is determined by a Monte Carlo scheme taking into account the four transition probabilities at each position n along the chain.

The input in the transformer is thus expanded through L hidden layers in accord with attention range g for the g-quintuples centered at each position n on the chain and influencing the basin transition at position n. The transformer input is a sequence of basin assignments for each residue along the chain, and its output is the result of the sequence transduction resulting in the basin assignment after time τ has elapsed. To apply the attention operation, a modulo-basin state for a chain of length N may be labeled with a 4N-sequence ($\mathbf{b}_1|\mathbf{b}_2|...|\mathbf{b}_n...$) consisting of N binary 4-tuples $\mathbf{b}_n$ ($n = 1, 2,...$, N) indicating the Ramachandran basin occupancy for each residue, so the value 1 at entry m in $\boldsymbol{b}_n$ ($b_{nm} = 1$) indicates that residue n occupies basin m ($m = 1, 2, 3, 4$) (Figure 3.1). For example, (1000|0010) indicates the coarse-grained state of a dipeptide with first residue in basin 1 and second residue in basin 3. Thus, the attention contribution has associated a convolutional kernel $K(n, g, B_K(n, g))$ that becomes a g-quintuple of binary 4-tuples (basin assignment $B_K(n, g)$) with range g, with the convolution operation carrying the Kronecker-delta factor $\delta_{A(n, g), A_K(n, g)}$, where $A_K(n, g)$ is the kernel amino acid assignment for positions $n - g - 1$, $n - g$, n, $n + g$, $n + g + 1$. Each sublayer $L(q, g)$ is labeled by the basin quintuple, in turn labeled by index $q = 1,...$, $4^5 = 1024$, and attention gap g. In each sublayer, the convolution kernels are slid along the chain as stencils, and full coincidence at position n in $L(q, g)$ yields concatenation (enrichment) at position n in the (q, g)-feature map with the four attributes $f(n, g, b \rightarrow b')$ for the four a priori possible b' destiny basins.

After processing across all L hidden layers, we arrive at a full representation of the chain at layer L, where each position n is endowed with L × 4 transition parameters $f(n, g, b \to b')$, one for each attention gap g and each destiny basin b'. In the L + 1 layer, the products $\Pi_{g=1}^{L} w(n, g, B(t)) f(n, g, b \to b')$ get computed, and the four transition probabilities $p_{ML}^{(t)}(n, b \to b')$ are obtained for each of the four a priori destinies b'. Finally, at the output, the destiny basin for each residue after time τ has elapsed from its original basin occupancy at time t is chosen from the Monte Carlo scheme.

3.2.2 Topological Metamodeling Requires Two Autoencoders and a Transformer

In accord with the notation and conceptual framework delineated in Chapter 2, the architecture of the underlying neural network that enables the topological metamodel discovery consists of two autoencoders, $AE1$: (γ, μ) and $AE2$: $(\pi, \mu^\#)$, and the transformer that generates the map Γ (Figure 3.2(a)). Thus, autoencoder AE1 discovers the latent manifold Ω of torsional dihedral coordinates representing the relevant internal degrees of freedom of the chain, whereas AE2 discovers the latent quotient manifold $\Omega/\sim$ of "modulo basin" classes that provides the topological "textual"

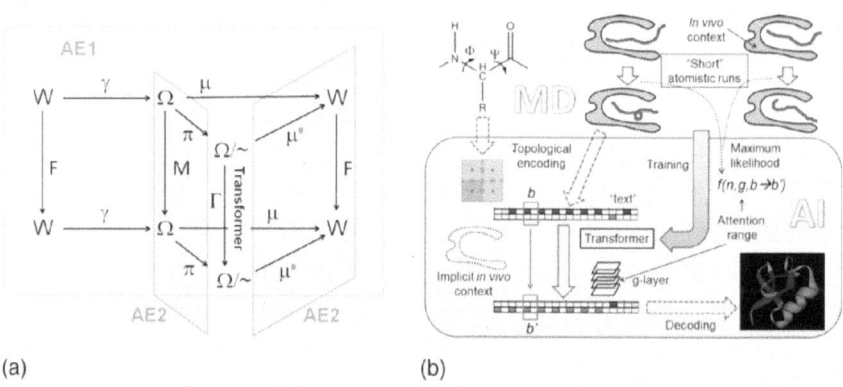

(a)   (b)

FIGURE 3.2 Neural network architecture and task flow for textual processing/ propagation of the topological dynamics metamodel of *in vivo* protein folding. (a) Neural network architecture for topological model discovery, consisting of two autoencoders, AE1 and AE2, and a transformer that generates the topological dynamics map Γ in the latent quotient manifold. (b) Task flow for a transformer capable of propagating short atomistic MD runs to generate the *in vivo* folding of a protein chain as "text processing." The "text" is the sequence indicating assignment of a Ramachandran basin to each residue along the chain. The alphabet consists of $20 \times 4 = 80$ "letters," each specifying amino acid type and basin occupancy.

framework of the metamodel. The autoencoder AE2 is not optimized for the topological dynamics subsumed in the map $\Gamma: \Omega/\sim \ \rightarrow \Omega/\sim$. This map is obtained by training and optimizing the transformer described in the previous section. The variational optimization of the transformer leads to the fulfillment of the commutativity relations: $\mu^{\#} \circ \Gamma = F \circ \mu^{\#}$ and $\pi \circ M = \Gamma \circ \pi$, which make it dynamically compatible with AE1 and AE2, as shown in Figure 3.2(a).

The task flow for the transformer network capable of computing the *in vivo* folding of a protein chain through text processing is described schematically in Figure 3.2(b).

The optimization of the attention-defining weights $w(n, g, B(t))$ is carried out during the second training phase for the modulo-basin propagator Γ_τ defined by the Monte Carlo generator of basin transitions. To train the transformer in this second stage, we break up the MD trajectories spanning a time period $[0, t_f]$ into two regions, a training portion with timespan $[0, t_0]$ and an optimization portion covering the time interval $(t_0, t_f]$. With the trajectory statistics drawn for the region $[0, t_0]$, we obtain the basin-transition frequencies for triads with different stride centered at each particular residue. On the other hand, a variational optimization process during the period $(t_0, t_f]$ enables us to determine the w-parameters. Thus, the ws are obtained from a commutativity condition that reflects the compatibility of fine and coarse-grained propagation of the trajectory and must be fulfilled during the period $(t_0, t_f]$. The commutativity condition is $\Gamma_\tau \circ \pi = \pi \circ M(\tau)$ as depicted by the following scheme:

$$x(t) \xrightarrow{\pi} \overline{x(t)}$$
$$\downarrow M(\tau) \quad \downarrow \Gamma_\tau$$
$$x(t+\tau) \xrightarrow{\pi} \overline{x(t+\tau)}$$

Here, π is the canonical projection – effectively, the encoding – that assigns the Ramachandran basins to each backbone torsional state of the chain (Figure 3.1) and $M(\tau)$ is the molecular dynamics propagator of torsional states of the chain. In other words, the parametrization of the modulo-basin propagator Γ_τ is optimized to minimize the loss function $\mathcal{L}(\Gamma_\tau)$ given by

$$\mathcal{L}(\Gamma_\tau) = M^{-1} \sum_{q=1}^{M} \|\pi x(t_0 + q\tau) - [\Gamma_\tau]^q \pi x(t_0)\|^2, \qquad (3.2)$$

where $M\tau \leq t_f - t_0 < (M+1)\tau$. The choice of loss function is correct since $\mathcal{L}(\Gamma_\tau) = 0$ if and only if the propagator makes the diagram commutative. Once the optimal $\Gamma_\tau^* = argmin\ \mathcal{L}(\Gamma_\tau)$ has been obtained from stochastic gradient descent, the modulo-basin projected trajectory can be propagated beyond MD-accessible timescales. The set $\wp(t_0, t_f)$ of MD-generated states during the period $t_0 \leq t \leq t_f$ is sampled randomly at each iteration to compute the gradient with minibatches given by trajectory fragments spanned by randomly chosen time intervals of size $10^{-2} |t_f - t_0|$. To optimize the network connectivity that determines Γ_τ, the ADADELTA optimization protocol with learning rate 0.3 is adopted [19]. Physically meaningful timescales may be reached by computing the coarse states $x(t_f + q\tau)$ beyond the simulation timespan t_f as $\left[\Gamma_\tau^*\right]^q x(t_f)$.

3.3 PROTEIN FOLDING AS A TEXTUALLY ENCODABLE DYNAMICAL METAMODEL: MATHEMATICAL VALIDATION

We now show that the protein folding dynamics may be reduced to an encodable dynamical system (EDS), characterized by an asymptotic behavior that is topologically described by a finite number of bits of information. To represent the protein chain dynamics as an EDS within a transformer platform (cf. Figure 3.2), it is necessary to prove that we can provide a textual representation of the enslaving modes that describe the backbone conformation dynamics (Figure 3.1) and subsume the determinant topological features of the force field [20]. An example of textual display of the backbone (Φ,Ψ)-representation would be a coarse-grained discretization that defines for each residue the local topology of the chain [9, 16].

For chain length N, conformation space Ω becomes the Cartesian product of N differentiable compact manifolds: $\Omega = \Pi_{n=1}^{N} \Omega_n$, where Ω_n is a 2-torus spanned by the dihedral (Φ_n, Ψ_n)-coordinates of the nth residue. Let $\mathcal{M} = \mathcal{M}(\{\Phi_n, \Psi_n\}_{n=1,\dots,N}) = -\nabla U(\{\Phi_n, \Psi_n\}_{n=1,\dots,N})$ denote the vector field defined on Ω and associated with the force-field potential energy U that governs the MD simulation of the folding process. The trajectories defined by the vector field $\mathcal{M}$ are τ-incrementally determined by $M(\tau)$, the propagator of torsional dynamics defined in Section 3.2. Let $\pi_n \mathcal{M} = \mathcal{M}_n : \Omega_n \to T\Omega_n$

denote the pseudo-projection of $\mathcal{M}$ onto Ω_n, with $T\Omega_n$ denoting its tangent bundle. The map $\pi_n \mathcal{M} = \mathcal{M}_n$ assigns to each point $y = (\Phi_n, \Psi_n)$ in Ω_n the projections onto $T\Omega_n$ of all the vectors $\mathcal{M}(z)$ in $T\Omega$ associated with all points $z = (\{\Phi_{n'}, \Psi_{n'}\}_{n' = 1,\ldots, N})$ such that $(\Phi_{n'}, \Psi_{n'}) = (\Phi_n, \Psi_n)$ for $n' = n$, that is, all the points z in Ω that project onto y. In general, this pseudo projection is a multivalued map. To define a true projection, we define the blueprint of $\mathcal{M}_n$ on the local Ramachandran potential energy function $U_{R,n}: \Omega_n \to R$ (R = real numbers) as $\mathcal{M}_{R,n}(x) = \{\int[\delta(\alpha)\mathcal{M}_{n,\alpha}(x)]d\alpha\}\hat{e}(x)$, where $\hat{e}(x) = -\nabla_n U_{R,n}(x)/\|\nabla_n U_{R,n}(x)\|$ if $\nabla_n U_{R,n}(x) \neq 0$, $\hat{e}(x) = 0$, otherwise, $\alpha = argcos\left[\widehat{\mathcal{M}_{n,\alpha}(x).\hat{e}(x)}\right]$ for $\hat{e}(x) \neq 0$, and $\delta(\alpha)$ is the Dirac delta.

We now introduce the following:

Definition. A smooth (i.e., class C^1, with continuous first derivative) vector field V defined on a differentiable compact manifold is *textually encodable* (cf. Figure 3.1) if it can be approximated arbitrarily closely by a C^1 vector field W topologically determined by the basins of attraction of a finite set $x_1, x_2, \ldots, x_J$ of generic singular points. A singular point x is generic if $W(x) = 0$ and $Re\lambda(x) \neq 0$, for all λ = eigenvalue of the Jacobian matrix of W at x, and the basin of attraction of a singular point x_i is the set of points x whose destiny or omega set ($\omega(x)$) along the trajectory or integral line defined by W is x_i ($\omega_V(x) = x_i$). The distance $d(V, W)$ between two vector fields V and W is defined via $d(V, W) = \int \|V(x) - W(x)\|^2 dx$ where integration extends over the domain manifold common to both fields.

According to the definition of textually encodable, the flow defined by V can be approximated arbitrarily closely by a flow W coarsely defined by transitions between basins of attraction of J singular generic points. This implies that the flow V becomes encodable as text, that is, the generic point x may be represented by a binary vector $\bar{x}$ that simply specifies the basin of attraction relative to W that contains x.

To enable transformer technology to extend the molecular dynamics of unassisted and chaperone-assisted protein folding (Figure 3.2), we need to prove the following:

Theorem 3.1

The Ramachandran vector fields $\mathcal{M}_{R,n}(x)$ ($n = 1,\ldots, N$) are textually encodable, and hence the propagator (flow) $M(\tau) = \exp(\tau \mathcal{M})$ represents an EDS.

Proof. Since Ω_n is compact (a 2-torus), we simply need to prove that $\mathcal{M}_{R,n}$ can be approximated arbitrarily closely by a vector field Y with generic singularities (saddles, sinks, and sources). If they are generic, they are isolated by definition, and hence, since they do not accumulate, they must be finite in number. Since the singularities are finite, the vector field $\mathcal{M}_{R,n}$ becomes textually encodable, since the coarse-graining of Y only requires that we provide the spatial organization of the generic singularities of Y and the partition of Ω_n into a finite number of basins of attraction of sinks (2-dimensional) and saddles (1-dimensional, known as separatrices). An illustration of a textual encoding of the Ramachandran map is given in Figure 3.1.

Essentially we need to prove that for a given arbitrarily small $\varepsilon > 0$ we can find a vector field Y with a finite number of generic singularities satisfying: $\int \|\mathcal{M}_{R,n}(x) - Y(x)\|^2 dx < \varepsilon$, where integration extends over Ω_n.

The vector field $\mathcal{M}_{R,n} : \Omega_n \to T\Omega_n$ is a differentiable cross section of the tangent bundle $T\Omega_n$. The mapping $\mathcal{M}_{R,n}$ can be approximated by a map $L : \Omega_n \to T\Omega_n$, with L satisfying two conditions [21]: (a) $\int \|\mathcal{M}_{R,n}(x) - L(x)\|^2 dx < \varepsilon/2$ and (b) L is transversal to Ω_n, meaning that at each singular point of L, the Jacobian matrix is nonsingular ($\lambda \neq 0$), or the singularities of L are simple. The existence of L is guaranteed by Thom's transversality theorem [21], which posits that transversal maps are dense in the space of differentiable vector fields on a compact manifold and the observation that if a map contains simple singularities, there must be a finite number of them because (a) Ω_n is compact and (b) simple singularities are by definition isolated. By means of a small C^1-perturbation with norm $<\varepsilon/2$, we can turn L into the vector field Y with the desired following properties:

a. Y has a finite number singularities, all generic, and

b. $\int \|\mathcal{M}_{R,n}(x) - Y(x)\|^2 dx \leq \int \|\mathcal{M}_{R,n}(x) - L(x)\|^2 dx + \int \|L(x) - Y(x)\|^2 dx < \varepsilon$

As shown subsequently, this approximation theorem ensures applicability of the transformer platform to extend molecular dynamics spanning realistic timescales in accord with the scheme given in Figures 3.1 and 3.2.

We have shown that for any n, the Ramachandran map $\mathcal{M}_{R,n}$ is the accumulation point of a sequence $\{Y_j\}_{j=1,\ldots}$ of "simple" vector fields with a finite number of generic singularities on Ω_n. The modulo-basin quotient space $\dfrac{\Omega_n}{\left[\sim\right]_Y} = \Omega_n / Y$ with respect to vector field Y is defined by the map

$x \to \omega_Y(x)$ [22, 23]. As we approximate $\mathcal{M}_{R,n}$, there is a j-value $j = j^*$ beyond which the quotient space remains essentially invariant (if it would change, we would deviate from approaching $\mathcal{M}_{R,n}$). In other words: $\forall n \exists j^*(n): \dfrac{\Omega_n}{Y_j} \approx \dfrac{\Omega_n}{Y_{j'}} \forall j > j' > j^*$, where $\approx$ denotes "isomorphic." Then, since for j large enough, each approximant vector field contains the same number K of generic singular points, we may textually encode the local state x of the chain by specifying the singular point x_k ($k = 1,\ldots, K$) that constitutes the omega set (point) of x in accord with the surjective map $x \to \omega_{Y_j}(x) = x_k \forall j > j^*$. Thus, a binary K-tuple with as many entries as generic singular points for Y_j ($j > j^*$) serves as textual encoding for the coarse-grained state $\bar{x}$, with 1 for the kth-entry of the encoding vector and 0 for all other entries.

3.4 INJECTING *IN VIVO* REALITY INTO THE TRANSFORMER-GENERATED METAMODEL

We now incorporate the role of *in vivo* settings in steering the protein chain along an expeditious folding pathway. Given computational limitations, atomistic detail on a folding trajectory in an *in vivo* context, say, in the chaperone chamber, is hardly feasible. For a typical single-domain protein (N>50), such computations are not likely to reveal the mechanism by which the cellular context is able to expedite folding. To generate coarse-grained *in vivo* folding trajectories, we implement a deep learning system that captures the expediency of the *in vivo* environment when trained with accessibly short (30 ns) MD simulations starting at unfolded or metastable states generated *in vitro*. We need to capture the means by which the *in vivo* setting removes kinetic traps by selectively disrupting misfolded states that would otherwise be susceptible to aggregation.

Let $\widetilde{\Gamma}_\tau$ denote the modulo-basin propagator that commutes with $\widetilde{M(\tau)}$, the MD operator that steers the chain torsional dynamics in the *in vivo* setting (in our illustration, the GroEL cavity). The propagator $\widetilde{\Gamma}_\tau$ is obtained by training the transformer with many short *in vivo* runs, and its domain $\mathcal{D}\left(\widetilde{\Gamma}_\tau\right)$ consists of basin assignments along the chain that undergo basin transitions (including retentions) when the chain explores conformations in the *in vivo* setting. The domain for Γ_τ^* is similarly defined *mutatis mutandis* in association with the *in vitro* setting.

To generate *in vivo* folding trajectories that span realistic timescales, a significant coverage is required of the influencing basin occupancies at positions flanking each residue along the chain placed in the *in vivo* environment. To that effect, we construct the propagator $\widehat{\Gamma_\tau}$ obtained by overwriting basin transitions yielded by Γ_τ^* in the domain $\mathcal{D}\left(\Gamma_\tau^*\right) \cap \mathcal{D}\left(\widetilde{\Gamma_\tau}\right)$, replacing them by the transitions dictated by $\widetilde{\Gamma_\tau}$. Thus, we get $\widehat{\Gamma_\tau} = \chi_{\mathcal{D}\left(\widetilde{\Gamma_\tau}\right)} \widetilde{\Gamma_\tau} \otimes \chi_{\mathcal{D}\left(\Gamma_\tau^*\right) \setminus \left[\mathcal{D}\left(\Gamma_\tau^*\right) \cap \mathcal{D}\left(\widetilde{\Gamma_\tau}\right)\right]} \Gamma_\tau^*$, where χ denotes characteristic function (1 on its support, 0 elsewhere), and $\mathcal{D}\left(\Gamma_\tau^*\right) \setminus \left[\mathcal{D}\left(\Gamma_\tau^*\right) \cap \mathcal{D}\left(\widetilde{\Gamma_\tau}\right)\right]$ denotes the complement of $\mathcal{D}\left(\Gamma_\tau^*\right) \cap \mathcal{D}\left(\widetilde{\Gamma_\tau}\right)$ in $\mathcal{D}\left(\Gamma_\tau^*\right)$. Thus, the *in vivo* propagator $\widehat{\Gamma_\tau}$ commutes with $\widetilde{M(\tau)}$ in $\pi^{-1}\left[\mathcal{D}\left(\widetilde{\Gamma_\tau}\right)\right]$ and with $M(\tau)$ in $\pi^{-1}\left\{\mathcal{D}\left(\Gamma_\tau^*\right) \setminus \left[\mathcal{D}\left(\Gamma_\tau^*\right) \cap \mathcal{D}\left(\widetilde{\Gamma_\tau}\right)\right]\right\}$ implying that $\widetilde{\Gamma_\tau}$ overwrites Γ_τ^* in $\mathcal{D}\left(\Gamma_\tau^*\right) \cap \mathcal{D}\left(\widetilde{\Gamma_\tau}\right)$.

3.5 PROPAGATING *IN VITRO* FOLDING PATHWAYS

The efficacy of the quotient space simplification is illustrated by computing folding pathways converging to native structures. We selected an N=57 chain known to fold autonomously in an *in vitro* setting: the thermophilic variant of the B1 domain of protein G from *Streptococcus* (PDB.1GB4). The thermophile was chosen over the wild type due to higher stability of the folded structure. Using the charmm package (free version of CHARMM) [3], we first generated a 220 μs-folding trajectory within the NPT (isothermal/isobaric, $T = 298$ K) ensemble [16]. The system was equilibrated at 300 K and 1 atm, and the runs were performed in the NPT ensemble with a Nosé–Hoover thermostat [24–26] and an MTK barostat [27], with the mass of all hydrogen atoms set at 4 amu, while the time step was fixed at 3.0 fs. The full MD trajectory encoded as the modulo basin version is shown in Figure 3.3(a), with selected $t_0 = 120$ μs as training parameter and $t_0 < t \le 220$ μs as learning period to optimize the propagator Γ_τ. The coarse-grained trajectory is then propagated up to $t = 7500$ μs (Figure 3.3).

We note that the final state is stable, prevailing since the time of its inception at $t = 3108$ μs discerned on Figure 3.3(b) (full trajectory in Appendix). Furthermore, the decoded final stable state at $t = 7500$ μs (Figures 3.3(b) and 3.4), corresponding to the eigenvector associated with eigenvalue 1 of the matrix Γ_τ^*, is topologically equivalent to the native crystallographic state (PDB.1GB4), revealing the same pattern of antiparallel and long-range parallel β-sheets with α-helix packed against the β-sheet motif. The backbone-atom RMSD is estimated at 1.88 Å. The decoding of modulo-basin states into torsionally specified conformations is obtained using a deep learning system previously described [16–18].

In the *in vitro* folding context, at least two metastable states can be spotted to be distinctively generated at 0.7 and 2.7 ms (Figure 3.3), corresponding to the chain conformations displayed in Figure 3.4(b and c), respectively.

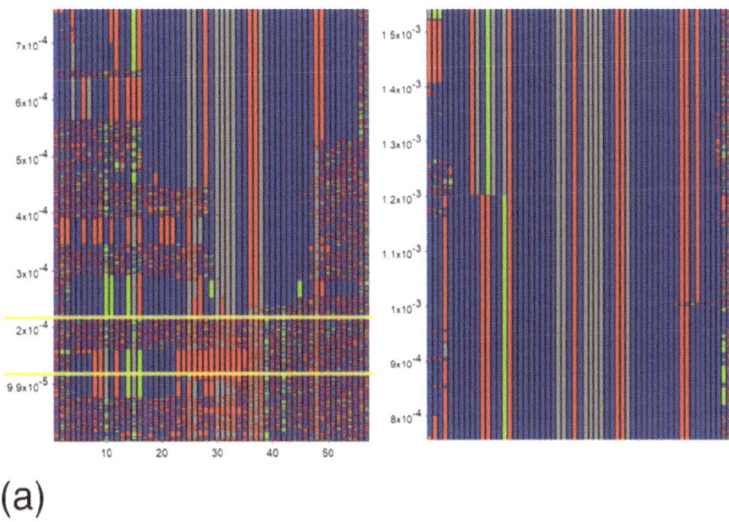

(a)

FIGURE 3.3 Coarse-grained propagation generated by a DL system of an atomistic MD trajectory covering 220 microseconds. Key portions of the MD trajectory at 1 µs-resolution covering timespan [0,7500 µs]. (a) First portion of the trajectory. The first 220 microseconds correspond to a coarse-grained representation of an MD trajectory at 1 µs-resolution for the N = 57 chain of the B1 domain of protein G. The trajectory is subsequently extended in time within a deep learning platform. At each time (seconds, vertical axis), the modulo-basin state of the chain is shown as a color sequence, where the basin assigned to each residue on the horizontal axis is specified in accord with the convention set forth in Figure 3.1. The MD trajectory was generated within an NPT (isothermal/isobaric, T = 298 K) ensemble, with interval [0, t_0=120 µs] as training region and interval [120 µs, 220 µs] as learning period.

(*Continued*)

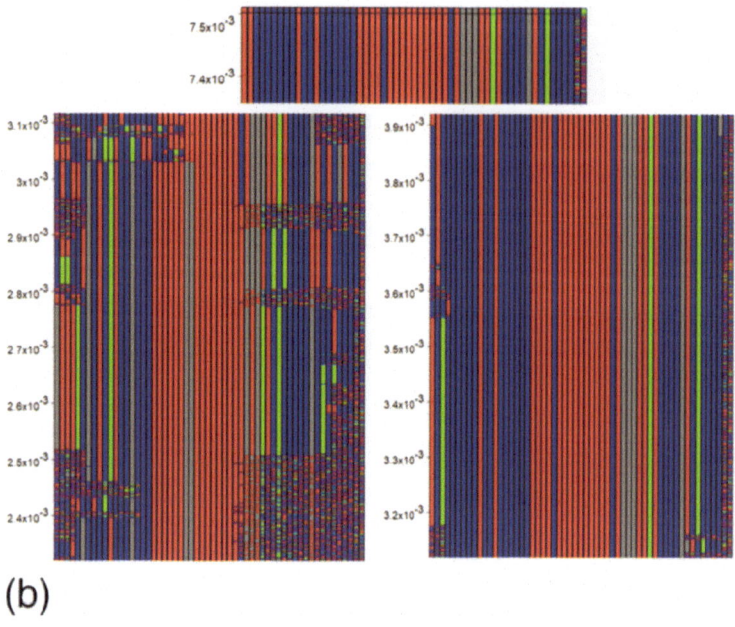

(b)

FIGURE 3.3 *(Continued)* (b) The steady-state ensemble is shown for the final portion of the trajectory (full trajectory in Appendix). The native-like steady state develops at around 3100 µs.

3.6 ATOMISTIC MD SIMULATION OF AN *IN VIVO* FOLDING SETTING

This section is purely methodological and describes a standard setting for molecular dynamics on a highly complex molecular reality. A conformation identified with the X-ray diffraction structure of GroEL(ATP)$_{14}$ (PDB.1SX3) was adopted to model the GroEL chamber assumed to expedite the folding of the B1 domain of G protein (PDB.1GB4). Excess Mg^{++}-complexed ATP molecules were removed, while the 22 inherently disordered C-terminal amino acids, mainly comprised of Gly-Gly-Met repeats, were incorporated, as previously specified [24]. One protein chain was placed at the center of the GroEL cavity, and the system was solvated in a 200 Å × 200 Å × 200 Å water box at 120 mM KCl concentration. The final system was composed of ~1,108,000 particles. Simulations of 30 ns were performed using the charmm package [3]. Seventy-two runs were performed with the substrate protein initially in the apo GroEL chamber with second chamber in (ATP)$_7$-state. The substrate in its resultant conformation was subsequently transferred as rigid body to the center of the (ATP)$_7$-chamber with the second chamber in (ADP)$_7$-state. Ten runs were

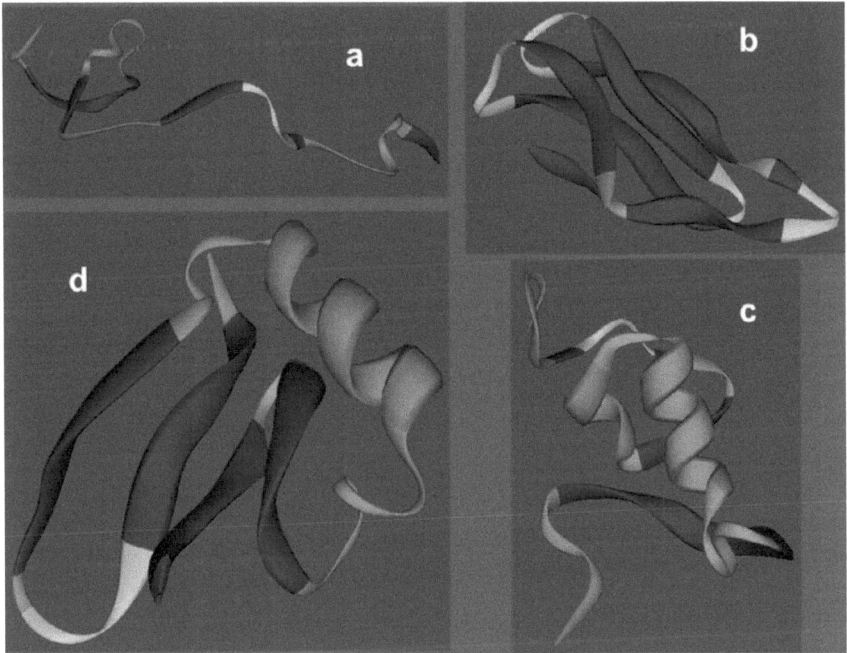

FIGURE 3.4 Decoded torsional states in ribbon representation for the AI-enabled propagation of the MD trajectory, generated at start (a), 0.7 ms (b), 2.7 ms (c), and 7.5 ms (d). The modulo-basin representation of the last microstate (d) is the eigenvector associated with eigenvalue 1 of matrix Γ_τ^* and is topologically equivalent to the native state from PDB entry 1GB4 obtained by X-ray diffraction crystallography, with RMSD = 1.88 Å.

started in an unfolded state, 2 in the protein native state (PDB.1GB4), and 60 were started in metastable states identified in *in vitro* runs, as illustrated below. The system was subject to energy minimization and equilibrated at 300 K and 1 atm with harmonic restraints on the alpha carbons. The runs were performed in the NPT ensemble with a Nosé–Hoover thermostat [25, 26] and an MTK barostat [27], with the mass of all hydrogen atoms set at 4 amu. The time step was fixed at 3.0 fs.

3.7 PROPAGATION OF *IN VIVO* FOLDING TRAJECTORIES USING TRANSFORMER TECHNOLOGY

To incorporate the influence of the *in vivo* reality, the transformer choreographs folding disruptions at states, where Γ_τ overwrites Γ_τ^* when the application of the latter generates metastable states characterized by the persistence of specific basin assignments. The transformer operationally incorporates the *in vivo* intervention step. Assume a substantial portion (typically 80% or more) of the basin assignment for the chain is detected to be locked at time $t_f + q_s\tau$, then the potentially disruptive propagator $\widetilde{\Gamma_\tau}$ overwrites Γ_τ^* for the step $t_f + q_s\tau \rightarrow t_f + (q_s + 1)\tau$. Thus, the coarse state at $t_f + (q_s + 1)\tau$ is computed as $\widetilde{\Gamma_\tau}\left[\Gamma_\tau^*\right]^{q_s} x(t_f)$. The disruptive propagator excludes the basin retention $b' = b$ frequency $f_\tau(n, b, b)$ from the Monte Carlo scheme that assigns at time $t_f + (q_s + 1)\tau$ the destiny basin b' at chain position n. For residues that did not retained their basin assignment at time $t_f + q_s\tau$, the propagation $b \rightarrow b'$ is constructed as in Γ_τ^*.

3.8 THE AI PLATFORM TO GENERATE *IN VIVO* FOLDING PATHWAYS

To illustrate the generation of *in vivo* folding pathways, the AI platform constructed an *in vivo* reality by learning to generate the propagator $\widehat{\Gamma_\tau}$ from 72 atomistic 30 ns-MD runs of the N=57 protein chain exploring conformation space within the apo and $(ATP)_7$ states of the GroEL chamber. The full catalytic round in the apo GroEL chamber is allowed to be completed before the chain in its final conformation is placed in the $(ATP)_7$ chamber, as indicated in Figure 3.5(a–c, f, g). The transformer-generated folding trajectory is shown in Figure 3.5(a), while Figure 3.5(b and c) show the influence of the *in vivo* context depicted as selective interferences with basin transitions. The basin assignments implicating the *in vitro* model valid in $\mathcal{D}\left(\Gamma_\tau^*\right)\setminus\left[\mathcal{D}\left(\Gamma_\tau^*\right)\cap\mathcal{D}\left(\widetilde{\Gamma_\tau}\right)\right]$ are marked in white, those involving *in vivo* intervention in $\mathcal{D}\left(\widetilde{\Gamma_\tau}\right)\setminus\left[\mathcal{D}\left(\Gamma_\tau^*\right)\cap\mathcal{D}\left(\widetilde{\Gamma_\tau}\right)\right]$ are marked in pink, while black denotes *in vivo* overwriting of *in vitro* assignments in $\mathcal{D}\left(\Gamma_\tau^*\right)\cap\mathcal{D}\left(\widetilde{\Gamma_\tau}\right)$ and gray indicates coincidence between *in vitro* and *in vivo* assignments in $\mathcal{D}\left(\Gamma_\tau^*\right)\cap\mathcal{D}\left(\widetilde{\Gamma_\tau}\right)$. The starting conformation is shown in ribbon rendering on Figure 3.5(d). The convergence to a native fold (Figure 3.5(e)), with RMSD = 1.91 Å relative to PDB.1GB4, occurs 75 times faster than in the *in vitro* setting (cf. Figure 3.3), attesting to the efficiency of the *in vivo* context to extricate the chain from kinetic traps during the annealing phase taking place in the apo state of the GroEL chamber.

Strikingly, ultra expeditious *in vivo* folding, 750 times faster than *in vitro* folding, are generated by AI with a transformer further enriched with *in vivo* reality (Figure 3.5(f and g)). This reality is extracted from 120 coarse runs spanning 100 μs each generated by the transformer trained as specified above with the MD runs subsuming the GroEL context.

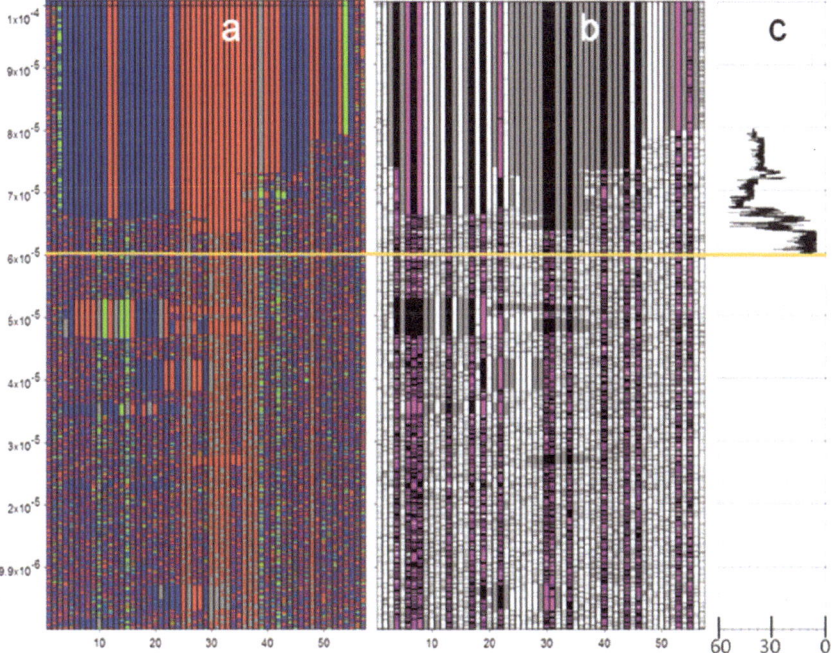

FIGURE 3.5 Expeditious *in vivo* folding trajectories generated in an AI transformer platform (Figure 3.2). The full catalytic round in the apo GroEL chamber is allowed to be completed before transferring the chain to the $(ATP)_7$ chamber. The time of transference is marked by the yellow line in Figure 3.5(a–c, f–g). (a) Coarse-grained trajectory at 0.1 μs resolution generated with a CNN trained with 72 atomistic 30 ns-MD runs of the chain dynamics within a GroEL environment. (b) Influence of *in vivo* context, as described in main text. (c) Burst of cooperative interventions of the *in vivo* context. Number of protein BHBs that are partially shielded from disruptive hydration by side-chain nonpolar groups in the flexible tails of the chaperonin, supplementing intramolecular BHB wrapping.

(Continued)

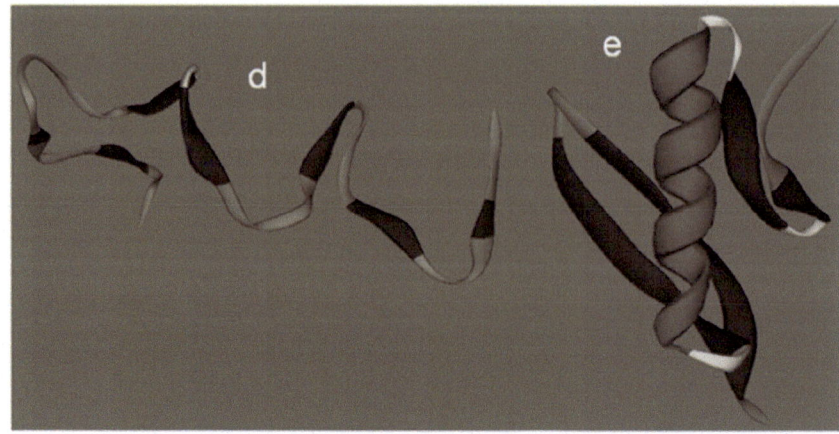

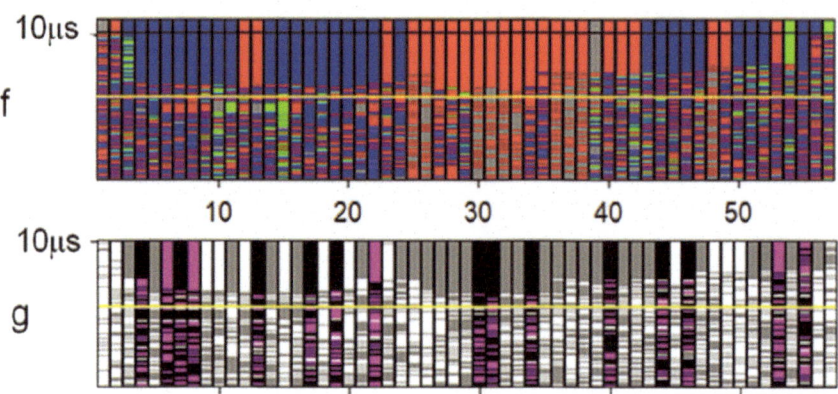

FIGURE 3.5 (*Continued*) Expeditious *in vivo* folding trajectories generated in an AI transformer platform (Figure 3.2). The full catalytic round in the apo GroEL chamber is allowed to be completed before transferring the chain to the $(ATP)_7$ chamber. The time of transference is marked by the yellow line in Figure 3.5(a–c, f–g). (d) Initial state. (e) Decoded final state. (f) Ultra expeditious *in vivo* folding trajectory at 0.1 μs resolution generated by "wiser AI" with a CNN further enriched with *in vivo* reality from 120 runs spanning 100 μs each and generated by the CNN trained as specified in (a). (g) Influence of *in vivo* context steering the ultra-expeditious folding pathway.

3.9 REVERSE-ENGINEERING THE EXPEDITIOUS *IN VIVO* CONTEXT I: ITERATIVE ANNEALING IN THE APO GROEL CHAMBER

The folding-assisting dynamics of the apo-state GroEL chamber consists basically of a tightly choreographed stochastic annealing process, which is very different from the folding assistance provided by the $(ATP)_7$ state of the

GroEL chamber. As the substrate is placed in the cage baricenter, the GroEL seven subunits are subject to conformational selection in both cases. Initially, in the apo state the T-conformation of the subunit, reported in PDB.1XCK, prevails for all seven subunits and constitutes a grabber of the folding substrate, as the C-terminus flexible hydrophobic tails comprising the last 22 residues of the subunit chain (526–548, absent in PDB structure) are initially free to interact intramolecularly. The T-centered ensemble occurs right after ADP is hydrolytically removed from the equatorial domains, a chemical event not explicitly modeled by the transformer. The T-centered ensemble with α-carbon RMSD dispersion at 2.2 Å for the initial 525 ordered residues prevails for ca. 4 µs and is denoted T_{grab} (Figure 3.6). However, after the substrate is tightly held in the chamber, two conformational ensembles named Grab and Rel (short for "release") become selected for the seven subunits. The Grab and Rel ensembles are centered respectively at PDB structures

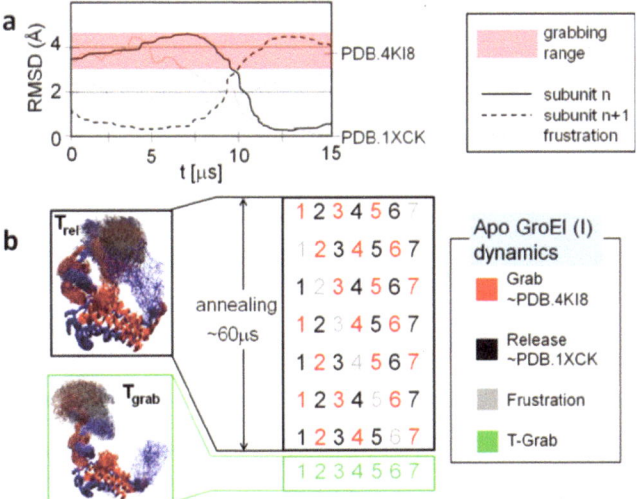

FIGURE 3.6 Dynamics of a catalytic folding-assistance cycle of the GroEL chamber in the apo state with subunit conformational selection resulting upon binding to the folding substrate (cf. Figure 3.5). Subunits are labeled by cyclic numerical indices (1, 2, …, 7, 8 = 1, 9 = 2, …). (a) Conformation dynamics reflecting allosteric anticorrelation across adjacent subunits (dark lines) and a pair including the frustrated subunit (gray lines). The RMSD is taken relative to PDB.1XCK. (b) Choreography of one complete annealing cycle determined by the alternating subunit roles of grabbing and releasing and the progressive displacement of the conformationally frustrated subunit along the annulus. The Grab (Rel) ensemble adopted by a subunit promotes transition to the Rel (Grab) ensemble in the adjacent subunit. Frustration is inevitable due to the odd number of subunits and their annular assemblage.

4KI8 ("R_{ADP}" state) and 1XCK (taut "T" state) with ensemble dispersions at 2.9 Å and 1.8 Å α-carbon RMSD, respectively. The substrate-grabbing role initially associated with the T_{grab} ensemble is now switched over to the R_{ADP}-centered ensemble Grab (Figure 3.6). Again, the inherently disordered and hydrophobic 526–548 tails containing the GGM-repeats and tethered to the C-termini of the equatorial domains are excluded from the RMSD calculations. In Grab, the C-terminus tails for subunits around the PDB.4KI8 conformation are free to interact with the substrate protein. By contrast, in Rel, the more rigid apical and equatorial domains interact intramolecularly through the hydrophobic C-terminus tails (Figure 3.6), which are therefore unable to bind the folding substrate and preclude capping of the chamber at the apical region by the GroES unit [28]. The two ensembles are allosterically anticorrelated (Figure 3.6(a)), so if one subunit is in Grab, the nearby unit transitions to Rel in a ~9.9 μs timescale and vice-versa, as shown in Figure 3.6(a). Because the number of anticorrelated subunits is odd (7), there is always conformational frustration (Figure 3.6(b)): one subunit tends to be in Grab and Rel simultaneously as it gets conflictive signals from its flanking subunits. This frustrated subunit slides through the ring (Figure 3.6(b)), with the mismatched structure with RMSD at roughly 1/2 the RMSD between Grab and Rel (Figure 3.6(a)). Even at an ensemble average level, the dynamics in the apo GroEL chamber reflects a breaking of the sevenfold symmetry at all times due to the frustration. The apo form "mechanically" dismantles misfolded kinetic traps, an assertion validated by contrasting Figures 3.3 and 3.5. This sort of intermittent annealing has been previously postulated in prescient kinetic models that fit experimental data [28] and now finds support at the molecular dynamics level.

The annealing interactions between the 526–534 C-terminus tails and the folding substrate are stochastic in nature and cyclically choreographed with alternating subunit participation to disrupt metastable states. Examples of such metastable states are shown in Figure 3.4(b and c). They constitute kinetic traps if the folding process were to take place in the test tube. With this analysis, the molecular underpinnings and mechanistic realization of the prescient "iterative annealing model" [28] are brought to light.

3.10 REVERSE-ENGINEERING THE *IN VIVO* CONTEXT II: GROEL CHAMBER IN THE (ATP)$_7$ STATE

By contrast with the GroEL chamber in the apo state, the (ATP)$_7$ chamber has only one conformational ensemble selected for its seven subunits upon incorporation of the folding substrate. This is a far more flexible and diverse ensemble than those observed for the apo chamber and is centered at PDB structure 1SVT (R-state), with maximum RMSD within the ensemble at 4Å for α-carbon resolution. This conformational selection is consistent with the previously established fact that the T→ R transition is triggered by ATP binding to the equatorial domain [28]. Using the flexible hydrophobic tails tethered at subunit C-termini, the $(ATP)_7$ chamber scaffolds folding-nucleating conformations by wrapping native-like backbone hydrogen bonds (Figure 3.5(c)), preventing structure disruption through backbone hydration.

Reverse-engineering the *in vivo* folding context is tantamount to elucidating how the *in vivo* environment cooperates with the folding process, making it expeditious. To accomplish this goal, we examined how budding native-like secondary structure that would not prevail in vitro becomes protected from disruptive backbone hydration in the $(ATP)_7$ chamber, enabling it to nucleate the formation of the native structure (Figure 3.5(c)). Since the main determinant of secondary structure is the backbone hydrogen bond (BHB), we decoded the modulo-basin representation into a tensor of three-body correlations (i, j, k), whereby a third residue (k) becomes wrapper of the BHB pairing residues i and j (Figure 3.7) [16, 18, 25]. By "wrapper" we mean capable of excluding water molecules that may otherwise form hydrogen bonds with BHB-paired residues i and j, thereby locally disrupting the protein structure. In practice, a residue k becomes an (i,j)-wrapper when it contributes side-chain nonpolar groups (CH_n, $n = 1, 2, 3$) located within a BHB microenvironment consisting of two intersecting balls of radius 6 Å centered at the α-carbons of the paired residues i, j [16, 18, 29, 30].

For proteins with reported structure, the extent of BHB wrapping, w, is identified from structural coordinates within the PyMol platform using a PyMol plugin [30] (code in Appendix). The local parameter w gives the number of side-chain nonpolar groups contained within a predetermined BHB environment. A batch-mode wrapping analysis of the PDB (download February 7, 2021) revealed that 86% of BHBs satisfy w = 26.6 ± 7.5, while all reported BHBs satisfy $w \geq 11.6$. The last relation indicates that sustainability of a BHB requires wrapping values higher than two standard deviations below the mean.

To reverse-engineer the *in vivo* context, we decode the modulo-basin state $x(t)$ into a wrapping (ijk)-tensor $L = L(x(t))$ (Figure 3.7) through adaptive learning. To that effect, we use a dilated CNN with weighted connectivity vector θ that infers BHBs and their respective wrapping in the

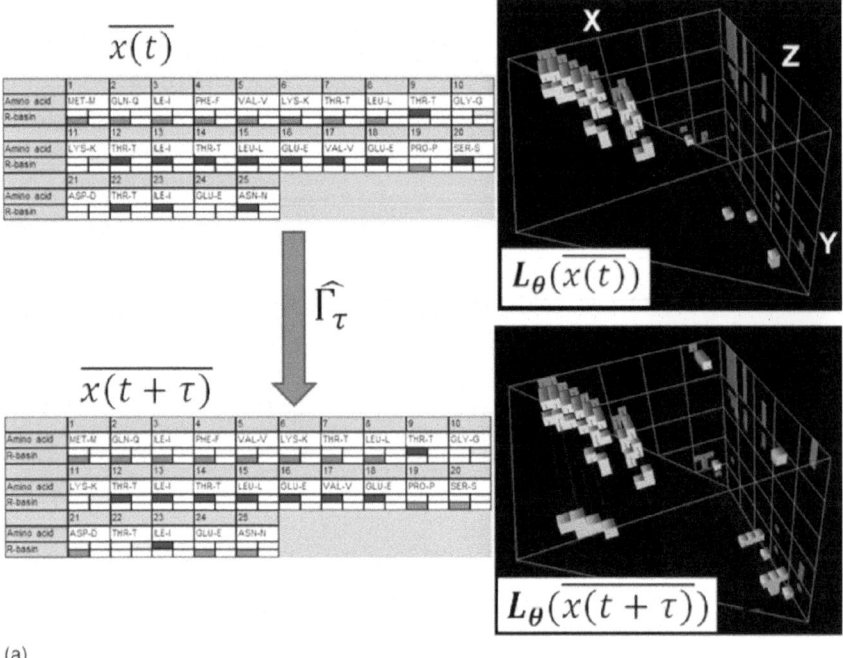

(a)

FIGURE 3.7 Decoding coarse "modulo-basin" states of the protein into wrapping tensors representing the extent of intramolecular shielding of BHBs from disruptive backbone hydration. (a) Change in $\overline{BHB}$ wrapping $L_\theta\left(\overline{x(t)}\right) \rightarrow L_\theta\left(\overline{x(t+\tau)}\right)$ corresponding to the *in vivo* transition $\overline{x(t)} \rightarrow \overline{x(t+\tau)}$ of a coarse-grained torsional state of a peptide backbone, specified by a Ramachandran basin assignment for each residue along a sequence of 25 amino acids. The wrapping tensor is represented as a tetrahedron, where the BHB (i,j)-pair (red voxel) is plotted in the X-Y plane and the wrapping residue generically denoted k (aqua voxel) is located along the Z axis. (b) The coarse-grained transition $\overline{x(t)} \rightarrow \overline{x(t+\tau)}$ corresponds to the molecular dynamics simulation of the torsional conformational change $x(t) \rightarrow x(t + \tau)$, which enables direct computation of the wrapping tensor transition $L(x(t)) \rightarrow L(x(t + \tau))$. (c) Feature extraction enabling the construction of the wrapping tensor $L_\theta\left(\overline{x}\right)$ from inputted coarse state $\overline{x}$. The hidden layers generate a torsional state of the chain that is compatible with $\overline{x}$ and draws the BHBs as gray ($w \geq 11.6$) or green ($w < 11.6$, vulnerable BHBs) segments joining α-carbons of BHB-paired residues. The wrapping of a BHB is represented by a blue line from the α-carbon of the wrapping residue to the center of the BHB.

(Continued)

(b)

$$x(t)$$

$$M(\tau)$$

$$\overline{x(t)}$$

$$\pi \rightarrow$$

$$x(t + \tau)$$

$$\widehat{\Gamma}_\tau$$

$$\overline{x(t + \tau)}$$

$$\pi \rightarrow$$

$$x(t + \tau)$$

	1	2	3	4	5	6	7	8	9	10
Amino acid	MET-M	GLN-Q	ILE-I	PHE-F	VAL-V	LYS-K	THR-T	LEU-L	THR-T	GLY-G
R-basin										
	11	**12**	**13**	**14**	**15**	**16**	**17**	**18**	**19**	**20**
Amino acid	LYS-K	THR-T	ILE-I	THR-T	LEU-L	GLU-E	VAL-V	GLU-E	PRO-P	SER-S
R-basin										
	21	**22**	**23**	**24**	**25**					
Amino acid	ASP-D	THR-T	ILE-I	GLU-E	ASN-N					
R-basin										

1-MET

25-ASN

(C)

FIGURE 3.7 (*Continued*) Decoding coarse "modulo-basin" states of the protein into wrapping tensors representing the extent of intramolecular shielding of BHBs from disruptive backbone hydration. (See previous caption for description of panels b and c).

feature-extraction stage [16, 29]. Feature extraction operates by generating torsional conformations compatible with $x(t)$ (Figure 3.7(c)) in layers filtered with receptive fields scaled by dilation parameters that correspond to contour distances ($|k\text{-}i|$, $|k\text{-}j|$) between wrapping residue (k) and BHB-paired residues (i, j).

As the flow progresses toward CNN output, the feature maps from hidden layers reveal longer and longer ranges of wrapping correlation. The decoding CNN adopts the loss function

$$ J(\theta) = M^{-1} \sum_{q=1}^{M} \left\| L\left(x\left(t_0 + q\tau\right)\right) - L_\theta\left(\left[\widehat{\Gamma_\tau}\right]^q \overline{x(t_0)}\right) \right\|^2, $$

where $M\tau \le t_f - t_0 < (M + 1)\tau$, $\|.\|$ is the Frobenius norm, and $L_\theta\left(\overline{x(t)}\right)$ is the inferred (learned) wrapping tensor associated with modulo-basin coarse state $x(t)$. Thus, the minimization of $J(\theta)$ is a least-squares problem numerically solved with stochastic gradient descent. To train the network, the ADADELTA method with learning rate 0.3 is adopted [19, 29].

Strikingly, during the nucleation period 60 μs $\le t \le$ 80 μs (Figure 3.5(a)), *none* of the BHBs in the wrapping tensor $L_\theta\left(x(t)\right)$ inferred by the optimized transformer satisfies $w \ge$ 11.6, making the nucleus thoroughly exposed to disruptive backbone hydration and therefore unstable unless exogenous BHB wrapping takes place. That means that the sustainability of the nucleating BHBs results from a burst of intermolecular protection of the protein conformations by the chaperonin during the structure-nucleating period 60 μs $\le t \le$ 80 μs (Figure 3.5(c)). This burst implies extensive exogenous contribution to the wrapping of the BHBs that eventually steer the formation of the native structure at $t >$ 80 μs. This finding suggests a nucleation mechanism that is unattainable in an *in vitro* setting due to insufficient wrapping or overexposure of the nucleus to disruptive hydration. *The GroEL chamber in the (ATP)₇ state intervenes cooperatively and stochastically to stabilize the folding nucleus, thereby committing the chain to fold.*

3.11 *IN VIVO* PATHWAYS FOR PROTEIN FOLDING: AI METAMODELS LIVE UP TO THE CHALLENGE

AI is revolutionizing molecular biophysics at a fast pace. Using AI, it has become possible to predict protein structure from amino acid sequence with staggering accuracy [1], the holy grail in the field a few years ago. One of the next frontiers involves predicting folding pathways and assessing the role of *in vivo* molecular contexts in expediting the folding process. This chapter represents a first step in that direction and shows that AI may fulfill the expectations through the construction of a metamodel. The effort is justified because natural proteins have been evolutionarily selected to fold in a cellular environment, not *in vitro*. It becomes imperative to learn how the cellular setting assists the folding process, prevents aberrant aggregation resulting from high local concentrations and disrupts metastable states that may constitute kinetic traps. This effort may broaden the technological base of the pharmacological industry since drug targeting efficacy ultimately needs to be assessed *in vivo*, not in the test tube, and drug-induced folding requires an assessment of time-dependent molecular processes, not merely of static assemblages [18, 29].

Computational efforts to generate *in vitro* folding trajectories are commendable and provide useful insights [8, 10, 11] whenever an Anfinsen scenario [7] can be upheld, but may arguably be misplaced in view of the fact that natural proteins have evolved to fold in an *in vivo* context. On the other hand, given the wanton complexity of cellular settings, the generation of *in vivo* folding pathways becomes unfeasible with current computational technologies. To address these challenges, this chapter introduced a transformer metamodel to recreate *in vivo* reality with the goal to reverse-engineer the molecular underpinnings of folding expediency. We built a transformer NN trained with numerous short (30 ns) MD runs to incrementally incorporate the very *in vivo* complexity that precluded MD from accessing realistic timescales and from capturing rare folding events. Atomistic MD simulations are unlikely to generate realistic *in vivo* folding pathways, even on dedicated supercomputers, but they can train an AI system such as the one presented in this chapter to enable the reverse engineering of cooperative *in vivo* contexts that expedite the folding process.

As an illustration, the transformer technology is deployed to reverse-engineer the steering participation of the ATP-consuming chaperone GroEL in the protein folding process [4–6]. We observe that the apo and $(ATP)_7$ states of the GroEL chamber play different roles as they contribute to expedite the folding process. The first acts stochastically to disrupt budding misfolds, with allosteric anticorrelated structural transitions between adjacent subunits taking place while a structural defect arising from conflictive flanking signals travels along the annulus and completes its cycle in about 60 μs. Thus, in the apo state, the GroEL chamber is shown to mechanically disrupt misfolded states in a cyclic choreography of alternating participating subunits. This choreography of alternating participants provides, at least to a degree, the molecular underpinnings to the prescient empirical model of "iterative annealing" [28]. On the other hand, the $(ATP)_7$ chamber participates stochastically in committing the chain to fold by protecting its folding nucleus. In this way, the GroEL $(ATP)_7$ state enables a productive initiation of the folding process that would be disrupted with nonnegligible probability due to competing backbone hydration if the folding process were to take place in bulk solvent.

3.12 METAMODELS WITH IMPLICIT CONTENT: CO-TRANSLATIONAL PROTEIN FOLDING

The participation of an *in vivo* context assisting the protein folding process may be encoded implicitly by adopting hierarchically coupled autoencoders, as described in Chapter 2. In this section, we turn to co-translational folding, whereby the budding peptide chain explores conformation space as it gets sequentially assembled in the ribosome, thus decoding information embossed in the messenger RNA. This decoding is known in biology as "translation." Thus, two generic states of the chain confined within their *in vivo* protein/RNA assemblages are regarded as equivalent under the relation "~" if the conformation of the chain is the same regardless of the conformation of the caging assemblages (Figure 2.12). Thus, both chain states map onto the same class modulo "~" under the canonical projection π. The autoencoder for the equivalence "~" is trained to fold the chain *in vivo* by treating the *in vivo* context implicitly. The training set consists of time series data generated by multiple molecular dynamics runs on the folding chain jointly with its confining molecular assemblage. None of the runs covers physically realistic timescales but taken together they instruct the autoencoder on how to implicitly encode information on the *in vivo* context.

A significant portion of natural proteins initiate and advance the folding process as they are being translated by the ribosome, and it is believed that about a third of the *E. coli* proteome is folded co-translationally, with comparable proportions for eukaryotes in general ([31] and references therein). The ribosome is a huge assemblage of protein and rRNA (ribosomal ribonucleic acid) subunits with total molecular weight of approximately 4.5 million daltons and endowed with mechanical, scaffolding, confining/caging, and enzymatic functionalities. The latter include peptidyl transferase, a tributary of the translational decoding of mRNA (messenger RNA) that performs an edit-prone sequential incorporation of

amino acids extending the nascent protein chain by catalyzing the formation of peptidic linkages [31]. The translation rate for eukaryotes is about
6 amino acids per second, while for prokaryotes that number increases to
about 20. Thus, it has been speculated that co-translational folding may be
a quasi-equilibrium process for those single-domain proteins that can
complete their *in vitro* folding in less than ~100 ms.

Most of the published results on co-translational protein folding (see,
e.g., [32]) remain mostly speculative or anecdotal. This is due mainly to
the forbidding complexities in modeling the ribosomal context. For start-
ers, with current computational capabilities, the molecular dynamics simulation of an assemblage with molecular weight on the order of 4.5 million
daltons is surely off limits, except for very short timescales on the order of
a few picoseconds. Thus, meaningful events involving the ribosomal pep-
tide-exit tunnel, the nanoscale-confined water therein and the ribosomal
processivity more broadly are out of reach with the standard tools of
molecular dynamics. The tunnel is a ribosomal region, 10–20 Å wide and
about 100 Å long, with rRNA lining (23S rRNA in bacteria, 28S rRNA in
eukaryotes), endowed with a constriction and ending in a wide vestibule.
These structural features are mainly formed by intercalating protein subunits, uL4/uL22 for the constriction, and uL23/uL24 for the vestibule,
with eL39 added in eukaryotes. We should highlight at this point that *residues in the vestibule are much less conserved across orthologs than residues
in other parts of the tunnel, an observation of key significance as we implement an AI system to unravel the process of co-translational folding.*

To assess the power of topological metamodels with implicit content, a
CNN with feature extraction system at the level of modulo-basin representation of the backbone torsional state of the growing chain has been
implemented as previously described. The goal is to capture the *in vivo*
participation in the co-translational folding of human ubiquitin
(UNIPROT ALIGN P0CG48), a single-domain protein with 76 amino
acids. The system was trained on thousands of short (1 μs) atomistic MD
runs with chains of different lengths ($4 < N \leq 76$) with the N terminus at
the peptidyl transferase center and started with all residues assigned to the
extended conformation basin to avoid steric hindrance. In addition, feature extraction was enhanced as the training set was enlarged to subsume
compatible coarse-grained structural and dynamic information from [31]
and references therein. Thereafter, atomistic trajectories for the budding
chains were propagated in the modulo-basin version as described in this
chapter, except that chain growth at the rate of $v = 6$ s^{-1} was incorporated

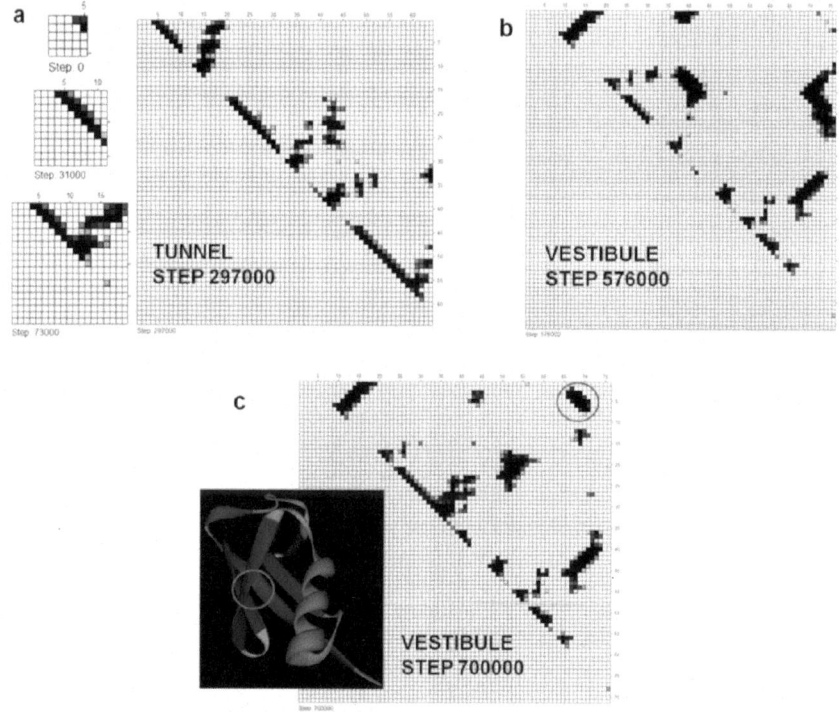

FIGURE 3.8 Contact matrix for the ubiquitin chain in the ribosomal peptide exit tunnel as obtained by AI-enabled propagation of a coarse-grained co-translational folding pathway. Snapshots at different stages of growth and folding (a–c). A ribbon rendering of the outcome structure (c) is given as an aid to the eye.

(Figure 3.8(a–c)). The CNN enabled inference of the co-translational folding pathway (Figure 3.8) with the full chain fully assembled in 355,000 steps corresponding to 12.66 s = N/v. Each coarse step covered 35.6 µs, and a steady-state conformation at α-carbon RMSD = 1.979 Å from the native fold (PDB.1UBI) was reached in 700,000 steps or 24.976 s (Figure 3.8(c)).

While at the tunnel, both native and nonnative secondary structures form (Figure 3.8(a and b)) but no scaffolding tertiary structure forms, so the secondary structure is ephemeral even in quasi-equilibrium conditions. Essentially, the tunnel prevents aggregation of misfolded structure that would otherwise take place as a result of the lack of long-range intra-molecular cooperativity. The intervention of the *in vivo* context is persistent at the N-terminus for obvious reasons and, strikingly, also near the

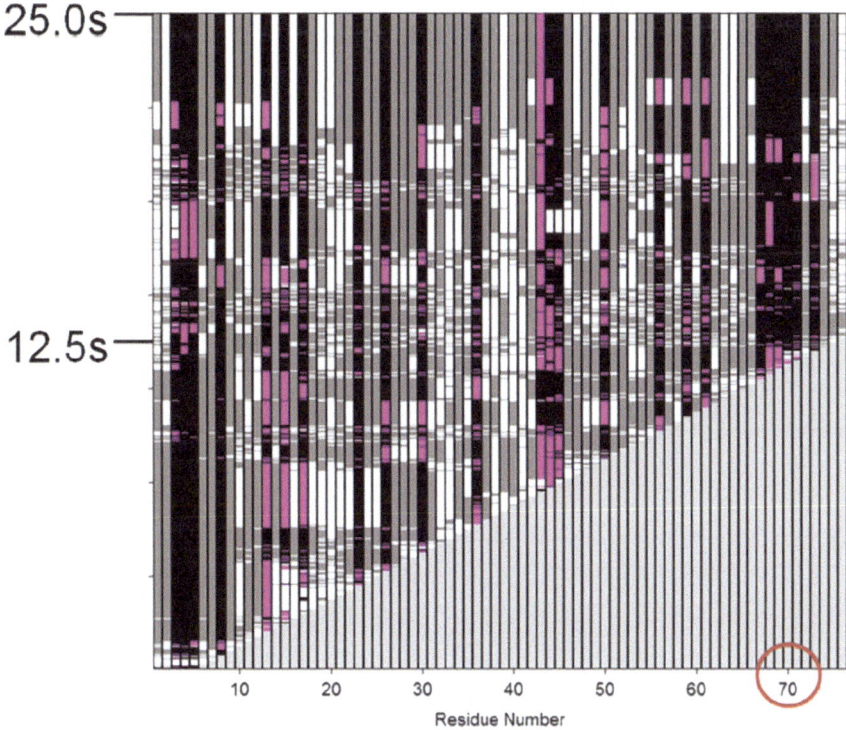

FIGURE 3.9 Influence of ribosomal context on the co-translational folding process assessed by AI. Depiction of *in vivo* interventions follows the convention described in Figure 3.5.

C-terminus, around residue 70 (Figure 3.9). This implies that at the vestibule, the C-terminus, the first region to exit the tunnel, interacts very strongly, specifically and persistently with the vestibular residues and that the long-range parallel *β*-sheet (circle in Figure 3.8(c)), which is nearly impossible to form *in vitro*, can form *in vivo* as an induced folding motif.

The ribosome influences the co-translational folding process so as to turn it into an induced folding process, implicating mostly the ribosomal proteins that shape the vestibule. The *in vivo* intervention, as captured by AI, then serves two purposes: (a) preclude aggregation of misfolded structure and (b) cooperate to induce the long-range structural organization (i.e., the parallel *β*-sheet in ubiquitin). While the induced unfolding has been mentioned in previous speculations [32], none mentioned the induced folding specifically implicating the vestibular ribosomal proteins (cf. [31]).

In this way, the *in vivo* complexity that expedites co-translational protein folding can be recreated by AI in order to shed light on an otherwise intractable setting that confers long-range organization to the native structure.

Shedding light on the molecular processes that take place within a molecular assemblage of about 4.5 million daltons is no minor feat, and is clearly off limits for any technology based solely on molecular dynamics. The enormous informational burden that needs to be carried over from one integration step to the next makes it absolutely forbidding to study co-translational folding with the standard tools of molecular biophysics. In this regard, the results shown in Figures 3.8 and 3.9 become iconic and epitomize the power and efficacy of AI-based metamodels to handle the wanton molecular complexities of *in vivo* reality.

This chapter shows that AI is capable of playing a transformative role in recreating and unraveling the molecular complexities of the cell and delineating their specific roles in assisting essential molecular processes that would probably not materialize otherwise or may do so in times which bear no relevance to life on earth.

REFERENCES

1. Callaway E (2021) DeepMind's AI Predicts Structures for a Vast Trove of Proteins. *Nature* 595:635.
2. Shaw DE, Maragakis P, Lindorff-Larsen K, Piana S, Dror RO, Eastwood MP, Bank JA, Jumper JM, Salmon JK, Shan Y, Wriggers W (2010) Atomic-Level Characterization of the Structural Dynamics of Proteins. *Science* 330:341–346.
3. Brooks BR, Brooks III CL, Mackerell AD, Nilsson L, Petrella RJ et al. (2009) CHARMM: The Biomolecular Simulation Program. *J Comp Chem* 30:1545–1615.
4. Clark PL, Elcock AH (2016) Molecular Chaperones: Providing a Safe Place to Weather a Midlife Protein-Folding Crisis. *Nature Struct Molec Biol* 23:621–623.
5. Thommen M, Holtkamp W, Rodnina MV (2017) Co-Translational Protein Folding: Progress and Methods. *Curr Opin Struct Biol* 42:83–89.
6. Sorokina I, Mushegian A (2018) Modeling Protein Folding in vivo. *Biology Direct* 13:13.
7. Anfinsen CB (1973) Principles That Govern the Folding of Protein Chains. *Science* 181:223–230.
8. Jiang F, Wu YD (2014) Folding of Fourteen Small Proteins with a Residue-Specific Force Field and Replica-Exchange Molecular Dynamics. *J Am Chem Soc* 136:9536–9539.

9. Fernández A (1999) Folding a Protein by Discretizing its Backbone Torsional Dynamics. *Phys Rev E* 59:5928–5934.
10. Englander SW, Mayne L (2017) The Case for Defined Protein Folding Pathways. *Proc Natl Acad Sci USA* 114:8253–8258.
11. Eaton WA, Wolynes PG (2017) Theory, Simulations, and Experiments Show That Proteins Fold by Multiple Pathways. *Proc Natl Acad Sci USA* 114:E9759–E9760.
12. Fernández A, Colubri A, Berry RS (2000) Topology to Geometry in Protein Folding: Beta-Lactoglobulin. *Proc Natl Acad Sci USA* 97:14062–14066.
13. Singhal N, Pande VS (2005) Error Analysis and Efficient Sampling in Markovian State Models for Molecular Dynamics. *J Chem Phys* 123:204909.
14. Zimmerman MI, Bowman GR (2015) Fast Conformational Searches by Balancing Exploration/Exploitation Trade-Offs. *J Chem Theory Comput* 11:5747–5757.
15. Ekman M (2021) *Learning deep learning: theory and practice of neural networks, computer vision, natural language processing, and transformers using TensorFlow.* Addison-Wesley, Boston, MA.
16. Fernández A (2021) *Artificial intelligence platform for molecular targeted therapy: A translational science approach.* World Scientific Publishing Co., Singapore.
17. Fernández A (2020) Deep Learning Unravels a Dynamic Hierarchy While Empowering Molecular Dynamics Simulations *Ann Phys (Berlin)* 532:1900526.
18. Fernández A (2020) Artificial Intelligence Teaches Drugs to Target Proteins by Tackling the Induced Folding Problem. *Mol Pharm (ACS)* 17:2761–2767.
19. Zeiler MD (2012) ADADELTA, an adaptive learning rate method. arXiv 2012, 1212.5701. https://arxiv.org/abs/1212.5701
20. Laskowski R, Furnham N, Thornton JM (2013) The Ramachandran plot and protein structure validation. In *Biomolecular forms and functions: a celebration of 50 years of the ramachandran map* (Bansal, M. & Srinivasan, N., eds.), pp. 62–75, World Scientific Publishing Co., Singapore.
21. Thom R (1954) Quelques propriétés globales des variétés différentiables. *Comment Math Helvet* 28:17–86.
22. Weisstein EW (2021) Quotient Space. From Math World--A Wolfram Web Resource. https://mathworld.wolfram.com/QuotientSpace.html
23. Nemytskii VV, Stepanov VV (2016) *Qualitative theory of differential equations.* Princeton University Press, Princeton, NJ.
24. Lippert RA, Predescu C, Ierardi DJ, Mackenzie KM, Eastwood MP, Dror RO, Shaw DE (2013) Accurate and Efficient Integration for Molecular Dynamics Simulations at Constant Temperature and Pressure. *J Chem Phys* 139:164106.
25. Nosé S (1984) A Unified Formulation of the Constant Temperature Molecular-Dynamics Methods. *J Chem Phys* 81:511–519.
26. Hoover WG (1985) Canonical Dynamics: Equilibrium Phase-Space Distributions. *Phys Rev A* 31:1695–1697.
27. Martyna GJ, Tobias DJ, Klein ML (1994) Constant Pressure Molecular Dynamics Algorithms. *J Chem Phys* 101:4177–4189.

28. Thirumalai D, Lorimer GH, Hyeon C (2020) Iterative Annealing Mechanism Explains the Functions of the GroEL and RNA Chaperones. *Protein Sci* 29:360–377.

29. Fernández A (2020) Artificial Intelligence Steering Molecular Therapy in the Absence of Information on Target Structure and Regulation. *J Chem Inf Model (ACS)* 60:460–466.

30. Martin O (2014) Wrappy: A dehydron calculator plugin for PyMOL. MIT License. http://www.pymolwiki.org/index.php/dehydron

31. Liutkute M, Samatova E, Rodnina MV (2020) Cotranslational Folding of Proteins on the Ribosome. *Biomolecules* 10:97.

32. O'Brien EP, Christodoulou J, Vendruscolo M, Dobson CM (2011) New Scenarios of Protein Folding Can Occur on the Ribosome. *J Am Chem Soc* 133:513–526.

## CHAPTER 4

The Drug-Induced Protein Folding Problem

Metamodels for Dynamic Targeting

To be is to be encoded.

– Ariel Fernández

4.1 PROTEIN STRUCTURE IS A DYNAMIC OBJECT: LESSON FOR TARGETED THERAPY

As described in Chapter 1, deep learning methods are making great strides toward solving the protein folding problem [1, 2]. In particular, a machine learning system named AlphaFold [1] (Deep Mind Technologies, UK) has proven most successful at predicting the 3D structure of autonomous protein folders based solely on sequence-derived input. AlphaFold fulfills a major imperative since elucidation of structure is essential to infer protein function, while the determination of protein structure quite often eludes the experimental methods available [1]. Yet, protein structure is not a static entity but a dynamic object, and, more often than not, proteins do not acquire their structure autonomously, but require binding partners to stabilize the fold. This observation has enormous consequences for drug discovery and molecular optimization, and force us to seriously reformulate the lock-and-key paradigm as the target may be part of a dynamic ensemble, not a single rigid structure.

DOI: 10.1201/9781003333012-6

In consonance with this argument, in this chapter we adapt the AlphaFold platform *mutatis-mutandis* to a different set of needs related to drug design, that is, we shall attempt to solve what we hereby name the *"drug-induced protein folding problem."* This task confronts us with the fact that target proteins are typically not fixed targets: they structurally adapt to the human-made ligand in ways that need to be predicted to empower pharmaceutical discovery. In stark contrast with AlphFold, which predicts structure for autonomous folders, we aim at predicting structures that only form and prevail within obligatory drug-target complexes, that is, structures for nonautonomous folders. The goal is to instruct drug design to target specific conformations that are a priori predicted to rely on association with the purposely designed drug to maintain their structural integrity. This is tantamount to adapt to AlphaFold platform to solve the drug-induced folding problem.

We can anticipate that AlphaFold per se cannot be put to good use in this context, simply because it does not incorporate 1D information that would signal regions susceptible to structural adaptation as the drug-target complex is formed. We need to incorporate disorder propensities in the 1D-input, and not just any flavor of disorder – for example, we don't care for regions that may represent merely a storage of conformational entropy – but rather disordered regions susceptible to turn into order upon binding. Metaphorically speaking, we are interested in incorporating signals for regions that represent "tame" as opposed to "wild" disorder.

To effectively incorporate these "tamed disorder signals" into an AlphaFold-derived platform, we first briefly revisit the AlphaFold architecture and workflow. AlphaFold trains with PDB-reported structures a multilayered convolutional neural network (CNN) [3], the architectural scaffold for feature extraction, and generates 3D structure models that fit accurate predictions of distances between residue pairs. Such predictions are based on information that quantifies pairwise evolutionary correlations [4] organized in a 2D array. The underlying premise for this representation is that the co-evolution of two residues within a multiple sequence alignment signals that they are spatially related. The co-evolution matrix elements are concatenated with residue profiling including amino acid identity and secondary structure prediction, *but no sequence-based prediction of disorder propensity*. The latter signals would be conflictive with overlapping secondary structure prediction and with the purview of the CNN. In fact, if disorder propensities are directly incorporated, they would clash along the feature-extraction phase in ways that would render the predictor ineffectual, as folding possibilities would be overwritten by disorder output or vice versa. Local propagation of distance constraints from one layer to the next is achieved via progressively dilated convolutional operations [5], where suitably chosen convolution kernels or "filters" allow incorporation of features not just from neighbors of a pixel but also from further afield, incorporating surrounding context by expanding the receptive field of the convolution stencils.

4.2 DEEP LEARNING TO TARGET MOVING TARGETS IN MOLECULAR THERAPY

Protein structure prediction assumes that the chain will adopt a 3D structure and that the structure is unique [1, 2]. Since AlphaFold is concerned with the endpoint and not with the folding process, it is immaterial whether or not the folding process can be successfully completed *in vitro*, that is, without the assistance of a cellular annealing, chaperone, and scaffolding apparatus. AlphaFold may well be successful at predicting structure of proteins that fold *in vivo* and are selected to fold only *in vivo*. On the other hand, molecular dynamics efforts toward elucidating the folding trajectories have been invariably misplaced because they have only dealt with the *in vitro* context, at least up to this day.

In practice, many proteins cannot fold autonomously and the structure is not unique but dependent on binding partnerships, reliant on such associations to acquire and maintain its integrity [6]. In fact, the binding context typically selects the protein fold from within an ensemble of closely related folding possibilities that we hereby name the induced folding ensemble (IFE). For example, proteins such as transcription factors typically possess only induced structure, that is, their fold cannot prevail in an "apo" (separated) form [7]; antibodies usually present antigen-induced conformational multiplicity [8], while proteins which constitute drug targets often have a variety of "holo" (within-complex) forms, depending on the ligand/drug they associate with [9]. These considerations imply that AlphFold by itself is ill equipped at dealing with folding context or folding-upon-binding scenario, which is precisely the scenario of interest to the drug designer. Thus, the pressing issue becomes, how do we modify or extend AlphaFold to encompass such possibilities?

The obvious signal that needs to be inputted in the DL platform for drug discovery built upon AlphaFold is the disorder propensity [7]. This is a sequence-based prediction of local propensity for disorder, a signal of the local conformational plasticity of the protein enshrined in a parameter named f_d, with $f_d = 0$ indicating certainty of order and $f_d = 1$ signaling certainty for disorder (Figure 4.1). As expected, f_d correlates with the wrapping parameter ρ described in Chapter 3. The latter parameter is an indicator of the level of local protection of the protein structure from the disruptive effect of backbone hydration. Thus, a local region that exposes the backbone is likely to relinquish or loosen its structure, and hence it is expected to have a higher disorder propensity. The marginally wrapped backbone hydrogen bonds are named *dehydrons* [7] and represent regions in the twilight zone between order and disorder (Figure 4.1) that may be predicted from sequence taking advantage of the f_d-ρ correlation (the computation of ρ is structure-based while the computation of its correlated parameter f_d is sequence-based). Thus, we may say that the disorder propensity is a dynamic signal that may be incorporated in a metamodel of induced folding.

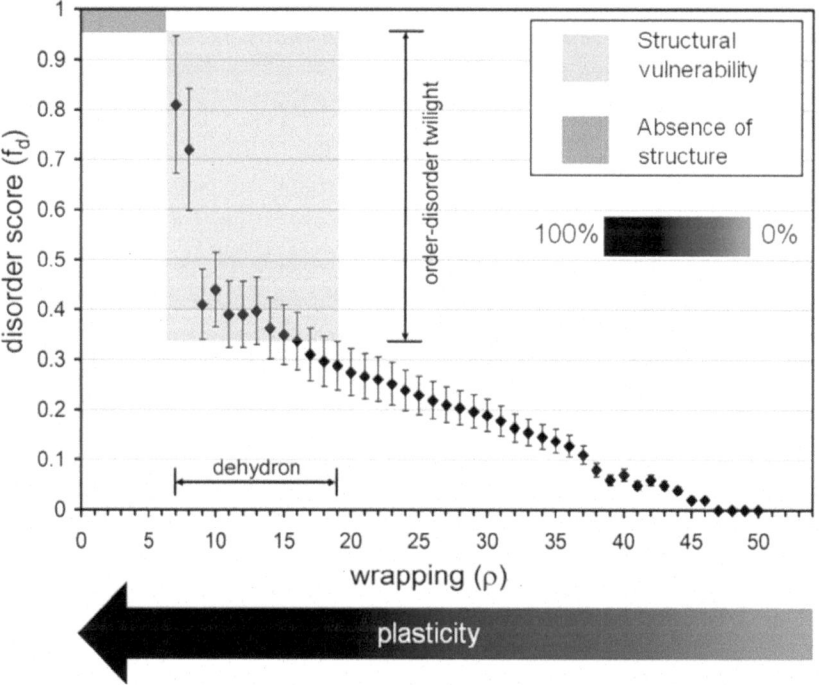

FIGURE 4.1 Correlation between sequence-based prediction of disorder propensity and structure-based computation of structure wrapping [7] in soluble proteins.

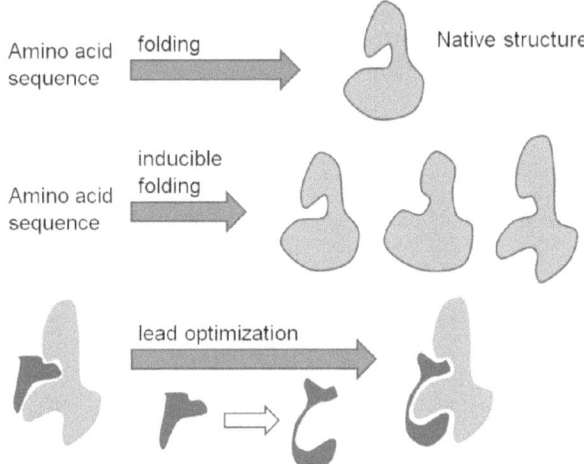

FIGURE 4.2   The drug-induced protein folding problem as differentiated from the protein folding problem solved by DeepMind. AI teaches the drug designer how to target a moving target and optimize the chemical scaffold to improve affinity.

The construction of a DL platform that incorporates dynamic signals to learn to target moving targets constitutes the focus of this chapter (Figure 4.2). The expected outcomes from this level of metamodeling may be illustrated by the structural diversity of the stress-responsive molecular chaperone Hsp90 [10], a cancer target that adopts different structures depending on the drug-target complex (Figure 4.3(a)). This type of induced folding diversity should prompt rational drug designers to focus on inferring the target's IFE, rather than a unique 3D structure, probably only suitable for autonomous folders [11] forming nonobligatory or ephemeral complexes. Thus, efforts in structure-based drug design are often marred by the need to encompass the target structural adaptation [8, 9]. Obviously, the IFE concept shifts the focus of structure prediction, especially of DL prediction, which must be adapted to infer a priori holo forms, introducing significant complications. Here, we propose a DL scheme to enable IFE prediction of pharmaceutically relevant holo forms, geared at tackling the "drug-induced folding problem," possibly the next frontier for a rational drug designer involved in lead optimization.

In this chapter, the AI predictions of IFEs are validated in two ways: (a) vis-à-vis experimental evidence on the conformation-selective power of AI-steered drug design and (b) through the generation of the predicted drug-induced conformation via an AI-empowered MD simulation that

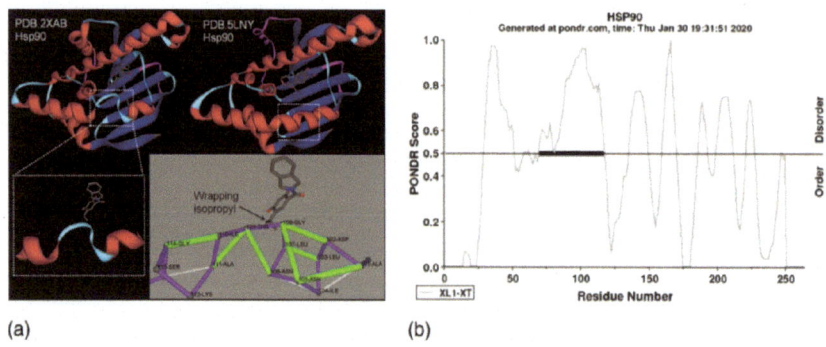

(a)   **(b)**

FIGURE 4.3   Conformational diversity of "*holo* forms" for cancer target Hsp90 induced by different drug ligands resulting from the intrinsic disorder propensity of a region in the protein, as the region is reliant on binding partnerships to maintain structural integrity. (a) The 100–120 helix in Hsp90 gets distorted, developing a kink as the ligand in the complex PDB.5LNY gets replaced for that in PDB.2XAB. The helical region has a disorder propensity which translates into susceptibility to fold into a binding-induced conformation that varies according to the ligand. The ligand in PDB.2XAB has a protruding isopropyl group capable of exogenously contributing to the wrapping of the dehydron-rich helix kink, specifically wrapping the dehydrons T109-A111 and N106-T109. In this way, the ligand determines or selects the binding-competent induced folding. BHBs are indicated as lines joining the α-carbons of the protein backbone, with well-wrapped bonds marked in gray and dehydrons in green. For the computation of BHB wrapping and identification of dehydrons, the reader may consult [12–14]. (b) Sequence-based prediction of intrinsic disorder for Hsp90 is regarded as an autonomous folder. The disorder plot was generated by the Predictor of Native Disorder (PONDR®), [15], freely accessible from http://www.pondr.com/. Reprinted from [Fernández A (2020) Perspective: Artificial intelligence teaches drugs to target proteins by tackling the induced folding problem". Molecular Pharmaceutics (ACS) 17:2761–2767] with permission from the American Chemical Society.

captures the structural adaptation of the target protein concurrent with formation of the drug/target complex.

4.3 AI-BASED METAMODEL TO INFER DRUG-INDUCED FOLDS IN TARGETED PROTEINS

The induced folding of the protein target implies that the protein relies on binding partnerships to acquire and maintain its structural integrity [6–9] with the ligand/drug selecting a conformation from within the IFE. Thus, delineating a suitable CNN architecture for the DL predictor of the IFE requires that we first identify the culprit for the lack of structural integrity of a free protein chain. We have learned from Chapter 3 that backbone hydrogen bonds (BHBs) are determinants of protein structure, and hence hydration of backbone amides and carbonyls competes with intramolecular hydrogen bond formation, exerting a structure-disruptive effect. An autonomous folder [11] is thus cable by itself of shielding or wrapping most BHBs from the disruptive effects of backbone hydration. Thus, protein integrity may be said to be contingent on the structure, by itself or within a complex, being capable of preventing backbone hydration. This leads us to propose wrapping as a descriptor of the degree of autonomy of the protein fold (or its degree of reliance on binding partnerships) [12, 13]. As it has been established, an underwrapped BHB constitutes a structural defect, known as dehydron [13], possessing an insufficient number of side-chain nonpolar groups clustered around the BHB, so that dehydrons are exposed to structure-disruptive hydration [6, 7, 12, 13]. An extended underwrapped region (dehydron cluster) has a high propensity to be disordered [7] in the *apo* form, since backbone hydration is likely to prevail over structure formation. Thus, induced folding may be interpreted as the optimal structural adaptation to a binding ligand to ensure the maximum shielding of underwrapped regions, thereby contributing to the stabilization of a specific *holo* form.

As shown in the introductory chapters, dehydrons and wrapping may be computed from structural coordinates of proteins [6, 7, 12, 13], and the computation may be incorporated in the feature-extraction workflow of the DL platform [14]. To that end, we introduce the extent of hydrogen-bond wrapping, ρ, indicative of the number of sidechain nonpolar groups contained within a "desolvation domain" that defines the BHB microenvironment in a training reported structure. Being marginally wrapped, dehydrons are located in the tail of the distribution of ρ-values across BHBs in the structural database. Dehydrons in protein structure compromise its integrity and promote protein associations as a means to exclude surrounding water (Figure 4.1). In this way, dehydrons become determinants for protein association because, by decreasing charge screening, exogenous removal of water from the dehydron microenvironment strengthens and stabilizes the electrostatic interaction that underlies the BHB. Ample bioinformatics evidence on the distribution of dehydrons at interfaces of protein complexes supports this picture [13], pointing to dehydrons as key factors driving complex formation, with dehydron-rich regions being flexible and hence susceptible of undergoing binding-induced folding [9]. To illustrate this picture, the Hsp90 helical 101–128 region in the complex reported in PDB entry 5LNY becomes disrupted when Hsp90 forms a different complex (PDB.2XAB), as shown in Figure 4.3(a). The structural difference arises as a kink forms with dehydrons N106-T109 and T109-A111, which in turn get exogenously wrapped by the protruding isopropyl group in the drug inhibitor that binds Hsp90 in the complex reported in PDB.2XAB. As the example suggests, the a priori inference of induced folding states would be essential for a rational drug designer, seeking to control drug affinity and specificity through lead optimization to select the *holo* form from within the IFE.

The previous discussion suggests that leveraging a DL platform to predict induced conformations hinges on a wrapping representation of the protein chain in the feature extraction phase. The structure wrapping must lend itself to a tensor representation wherein information is describable and processed on a multidimensional array (tensor) of digital entries within a layered architecture. Thus, the steering of the drug design process to target a binding-competent induced fold involves the inference of the IFE using as input the primary sequence of the target protein. This task requires an adaptation of the AlphaFold platform to incorporate the following elements:

1. In contrast with AlphaFold and other structure predictors, residues need to be profiled incorporating a key additional signal: the sequence-based prediction of intrinsic disorder, a descriptor of the propensity of a window along the chain to be inherently unstructured as the protein is taken in isolation, that is, as an autonomous folder [15, 16]. As shown in Figure 4.1, disorder score ranges from 1 (certainty of disorder) to zero (certainty of order). Regions predicted to be disordered are expected to be reliant on binding partnerships to improve the wrapping of BHBs to the point where they can be sustainable [6, 7]. *Thus, the disorder score signal needs to be integrated as a dynamic feature in the residue profiling for a sequence-based inference of the IFE.* A clear illustration is provided by the induced folding diversity in Hsp90: the predicted helicity of the region 100–120 (Figure 4.3(a)) is partially overlapping with an overwriting signal of intrinsic disorder for the region 80–114 (Figure 4.3(b)). This is a clear indication that the helix would be partially unsustainable if the protein is regarded as an autonomous folder, and can be partially disrupted or distorted by induced folding depending on the ligand, as it is indeed the case in the overlapping region 100–114 (Figure 4.3(a)).

Thus, a kink is induced in the predicted helical region, whereby nonhelical dehydrons N106-T109 and T109-A111 are formed and become sustainable as they get wrapped exogenously by the isopropyl group in the Hsp90-binding ligand reported in PDB.2XAB.

The overlapping conflictive signals described above inform us that the disorder score cannot be simply incorporated as input representation into the AlphaFold platform but requires an architectural modification of the underlying CNN to be adequately processed during the feature-extraction phase.

2. Residues paired by BHBs are represented within a triangular 2D array (the information plotted is symmetric) inputted with pairwise co-evolutionary information [3, 4]. The underlying premise is that if the evolution of two residues correlates across a multiple sequence alignment, it is likely that they are spatially related. The array is concatenated with residue profiling in the same way as it is done in AlphaFold, except for the incorporation of the local disorder propensity as an overwriting signal.

3. In contrast with AlphaFold, a third dimension is needed to identify the residues wrapping the BHBs that will emerge during the feature extraction phase of DL [13, 14]. Thus, wrapping components are identified upon the structural information within a 3D tensor whereby a matrix Y, evolving from the evolutionary coupling matrix (ECM), is constructed to ultimately indicate residues (i, j) paired by BHBs, while the wrapping of the BHBs is represented along Y-orthogonal vector z_{ij} for entry Y_{ij}. Thus, the DL flow operates on the tetrahedral $N \times N \times N$ tensor $Y \otimes Z = [Y_{ij} \otimes z_{ij}]_{ij}$ (N = chain length).

4. Feature extraction requires a dilated convolutional layered architecture [1, 3] where in a generic layer, denoted F, the convolutional kernel becomes a $3 \times 3 \times 3$ tetrahedral tensor (F) that evolves during the parameter optimization process, turning into a filter. This filter extracts a structural pattern drawn from the training set in a curated PDB-derived database [14]. The product of the convolutional operation $(Y \otimes Z) * F$ is thus a tensor with entries indicating sums of entry-by-entry multiplication (Frobenius inner product) of $Y \otimes Z$ by tensor F as the latter slides along $3 \times 3 \times 3$ tetrahedral sectors of $Y \otimes Z$ with stride 2 to produce the $(N-2)^3$ convolved tensor $\left[\widehat{Y \otimes Z} \right]_F = (Y \otimes Z) * F$. Thus, large value entries in the convolved

tensor are indicative of regions where the configuration more clearly resembles the kernel/filter. Within the F-layer, the convolved tensor $\lfloor \widehat{Y \otimes Z} \rfloor_F$ is expanded into the "F-feature-discerning" N^3-tensor $(Y \otimes Z)_F$ by spanning the F-pattern for the largest value entries in $\lfloor \widehat{Y \otimes Z} \rfloor_F$ while retaining the original entries from $Y \otimes Z$ for those $3 \times 3 \times 3$ tetrahedral sectors that don't yield a significant inner product. The F-filter is then dilated within an F-spanned multilayered architecture to incorporate surrounding context by enlarging the receptive field size of the convolution kernel. In this way, we incorporate features not just from pixel neighbors but from further afield by creating $3 \times 3 \times 3$ tetrahedral filters with the pixels further and further apart. Thus, each neuron in a hidden layer only receives input from the local region of what the input to the layer is, and with the dilations we enlarge the receptive field of the convolution kernel. So the hidden layer for filter F is actually spanning a bunch of hidden layers where the dilated filters $F_1 = F$, F_2, F_3, ... are successively applied, with the subindex giving the value of the dilation coefficient.

5) The training of the CNN enables optimization of the set of filters $\mathcal{F} = \{F^{(1)}, F^{(1)}_2 \ldots, F^{(2)}, F^{(2)}_2,...\}$ by minimizing the sum of Frobenius norms ($\|.\|$, sum of squared entries) for tensors $Y \otimes Z - \widehat{Y \otimes Z}$, where $Y \otimes Z$ indicates the structure-based computed wrapping of each BHB in a PDB-reported structure (ξ) picked from the training set $(\mathfrak{I})$, and $\widehat{Y \otimes Z}$ is the DL-based inference of the structure wrapping obtained from the protein sequence of the PDB-reported structure. Thus, the loss function $\mathcal{L}(\mathcal{F})$ for the DL network becomes

$$\mathcal{L}(\mathcal{F}) = |\mathfrak{I}|^{-1} \sum_{\xi \in \mathfrak{I}} \| (Y \otimes Z - \widehat{Y \otimes Z})(\xi) \|,$$

where $|\mathfrak{I}|$ denotes the number of structures in the training set.

While the tensor representation of structure wrapping allows suitable processing in a DL platform with Tensor Flow [17], the tensor should get decoded into a 3D rendering specifying relative spatial locations of wrapping residues vis-à-vis the BHBs to enable the threading of structural decoys onto it (Figure 4.4). The threading requires a separate inference of the torsional (Φ, Ψ) conformation of the protein backbone, as performed

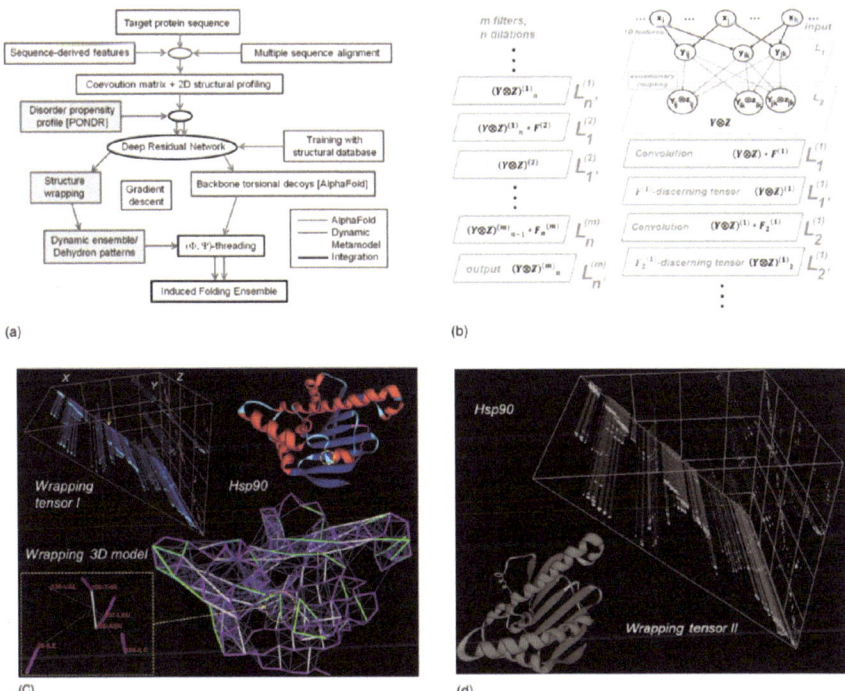

(a)

(b)

(c)

(d)

FIGURE 4.4   The DL discovery platform and its inference of the IFE for Hsp90. (a) Workflow for the DL system. (b) Architecture of the dilated convolutional network required to infer structural wrapping patterns. The dummy indices i, j, k indicate residues along the chain, and the 1D features, including disorder propensity, are subsumed vectorially in x_i, x_j, x_k, …. For feature extraction, the network contains $m \times n \times 2$ hidden layers labeled $L^{(1)}_1, \ldots, L^{(m)}_n$ (m = number of filters/convolution kernels, n = number of dilations per filter). In the cases reported in this study, we used $m = 120$, $n = 6$, so the network has 1,440 hidden layers. (c) Component I of the Hsp90 IFE, specified by the tetrahedral wrapping tensor, the decoded 3D wrapping model and the threaded 3D backbone conformation in ribbon rendering. The inferred BHBs are represented in the triangle in the background of the tetrahedron, while the residues contributing to the wrapping of each BHB are identified along the axis orthogonal to the background triangle. In the 3D wrapping model of the chain, a residue contributing to the wrapping of a BHB is indicated by a thin blue line between the α-carbon of the wrapper and the center of the wrapped BHB. (d) Component II of the IFE for Hsp90, specified by wrapping tensor and ribbon rendering of the 3D structure. Reprinted from [Fernández A (2020) Perspective: Artificial intelligence teaches drugs to target proteins by tackling the induced folding problem". Molecular Pharmaceutics (ACS) 17:2761–2767] with permission from the American Chemical Society.

by AlphaFold, and a fitting of the backbone torsional model onto the decoded 3D model of structure wrapping to determine side-chain locations to the level of resolution where it becomes possible to decide whether they contribute to wrap specific BHBs (Figure 4.4(c)). The full workflow of the DL platform is delineated in Figure 4.4(a).

4.4 LEARNING TO INDUCE FOLDS IN TARGETED PROTEINS

A DL system to solve the drug-induced protein folding problem has been constructed as described in Figure 4.4(a and b) following the premises 1–5 given in the previous section. Details for a basic precursor of this system were previously reported [14]. The precursor system has been modified *mutatis-mutandis* to infer induced folding possibilities. The repurposing requires the following: (a) The integration of a sequence-based prediction of disorder [15, 16] as a signal overwriting secondary structure prediction in the residue profiling. This signal is essential to identify regions reliant on binding partnerships to maintain structural integrity [7, 9], and hence susceptible of induced folding in consonance with the ligand. (b) The incorporation of dilated convolution in the layered architecture for feature extraction. This is essential to add nonlocal context to the prediction of induced folding, as the receptive field for each layer input gets progressively magnified in accord with the value of the dilation parameter for a given convolution kernel or filter during the feature extraction phase of the DL processing.

The IFE for cancer target Hsp90, obviously excluded from the training set and obtained by incorporating the prediction of intrinsic disorder given in Figure 4.3(b), is made up of two components, I and II, displayed in Figure 4.4(c and d). The display includes structure wrapping tensor and ribbon rendering of the (Φ, Ψ)-threaded conformation. The RMSDs of backbone Cartesian coordinates for I and II relative to the structural coordinates within the complexes (PDB.2XAB and PDB.5LNY, respectively) are 1.9 and 1.2 angstrom.

4.5 EXPERIMENTAL CORROBORATION OF THE DYNAMIC METAMODEL FOR DRUG-INDUCED FOLDING

Drug-based targeted therapy, aimed at blocking specific dysfunctional proteins, often faces a major obstacle due to induced folding, a hard-to-predict phenomenon that often generates unexpected and undesirable cross-reactivity while making the intended target elusive to molecular recognition [9]. We advocate that DL can steer drug design to achieve therapeutic impact by controlling the induced folding in the target protein. The way DL may teach the drug to target the protein is apparent as we focus on reworking the anticancer drug *imatinib* [18] into the prototype WBZ_4 ([9, 13], Figure 4.5). Exploiting wrapping differences among proteins that share a common fold as a filter for selectivity, a chemical modification of imatinib has been introduced to steer the drug impact toward clinically relevant imatinib targets (bcr-ABL, c-KIT, PDGFR kinases [19]), while suppressing off-target cross-reactivity against kinases such as LCK, whose inhibition may lead to harmful immunosuppressive effects [19]. However, just like we may remove potentially harmful cross-reactivity by chemically modifying the parental scaffold guided by the wrapping filter [18], we could also incorporate new and therapeutically desirable cross-reactivity guided by a kinome-wide examination of IFEs.

By inputting primary sequence, multiple sequence alignment, intrinsic disorder and secondary structure prediction, and pairwise residue coevolution, the IFEs for the 518 human kinases become accessible adopting the DL platform described. As expected, the augmentation of the imatinib scaffold by addition of a methyl group (WBZ_4 in Figure 4.5) would generally decrease cross-reactivity simply because less targets are typically able to accommodate a larger ligand as a result of steric hindrance. This is only partially true, however, as some targets are floppy enough to circumvent steric hindrance, especially those that may be susceptible to binding-induced folding. On the one hand, it was experimentally observed that WBZ_4 is not reactive against imatinib target ABL kinase or against LCK, culprits of cardiotoxicity [20] and immunosuppression [19], respectively. However, we found an exception to the steric hindrance rule in JNK1/2, which has no detectable affinity for imatinib but is a target for WBZ_4 [9]. JNK stands out in the kinome-wide IFE inference because of its significant induced folding diversity in the targeted region and its susceptibility to get wrapped by the extra methyl group not present in the imatinib scaffold, as shown in Figure 4.5. In fact, the nanomolar affinity of WBZ_4 against JNK1 has been experimentally documented [9], making this compound potentially impactful on ovarian cancer, where JNK inhibition has shown clinical relevance, at a variance with imatinib, which is not active in that therapeutic context [20].

The DL-inferred IFE for JNK1 has two components described in Figure 4.5: an apo form (I), where the dehydron BHB M111-N114 is not formed,

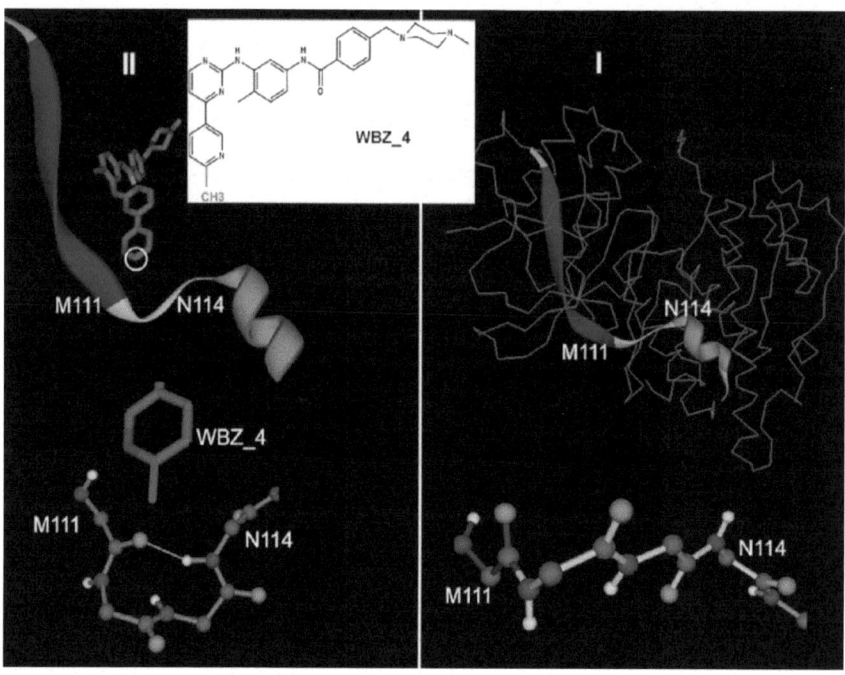

FIGURE 4.5 The IFE inferred by the DL platform for JNK, a target of WBZ_4 and not of parental drug *imatinib*. The *apo* form is component I and the *holo* form is II. The BHB M111-N114 (actually a dehydron) is induced upon binding to WBZ_4 and absent in the apo form. The two components have identical structure except for the region where induced folding occurs, indicated by the ribbon rendering. This example illustrates the application of AI to steer drug design in order to target a specific protein by learning to induce a drug-binding competent fold. Reprinted from [Fernández A (2020) Perspective: Artificial intelligence teaches drugs to target proteins by tackling the induced folding problem". Molecular Pharmaceutics (ACS) 17:2761–2767] with permission from the American Chemical Society.

and a holo form (II), which we reason must be the conformation adopted in the complex with WBZ_4. We arrived at that conclusion by docking WBZ_4 to JNK according to the structural alignment of JNK with the cKIT/imatinib complex reported in PDB.1T46 and noticing that WBZ_4 binding induces the BHB M111-N114 by contributing exogenously to its wrapping with the extra methyl that is not present in imatinib (Figure 4.5). Thus, the induced folding enabling the exogenous protection and hence stabilization of the induced M111-N114 BHB is responsible for the new cross-reactivity that gets turned on by tuning the parental chemical scaffold.

This example illustrates the use of the DL platform to steer the reworking of a drug in order to redirect its impact toward a different therapeutic context.

4.6 A TOPOLOGICAL METAMODEL OF THE INDUCED FOLDING DYNAMICS CORROBORATES THE EXPERIMENTALLY VALIDATED STRUCTURAL ADAPTATION IN THE DRUG/TARGET WBZ_4/JNK COMPLEX

To further validate the AI inference of the JNK conformation-selecting modification WBZ_4 of imatinib, we carried out the MD simulation of the WBZ_4/JNK complex formation in order to capture the structural adaptation of JNK to its putative ligand WBZ_4. To that end, we empowered MD using AI, along the lines described in Chapters 2 and 3.

To reduce the informational burden in MD integration steps, we simplify the dynamics taking into account the topology of the MD-steering vector field. Thus, microstates/conformations are regarded as equivalent if and only if the respective local conformations of the constitutive subunits lie in the same basins of attraction in the local potential energy surface. This is the "modulo basin" coarse representation of the protein conformation as described in Chapter 3. If x indicates a microstate of a system of N subunits, the modulo-basin class (state) to which x belongs is denoted $\bar{x}$ and may be expressed as a Cartesian product of N basins designating the basin occupancies for the individual subunits: $\bar{x} = \Pi_{n=1}^{N} b(x,n)$, where $b(x, n)$ is the basin of attraction occupied by subunit n when the system is in microstate x.

Folding and induced folding trajectories are defined by the evolution of backbone torsional coordinates $\{\phi_n, \psi_n\}_{n=1,...N}$. In this context, the modulo-basin "quotient" space is made up of Cartesian products of the so-called Ramachandran basins, that is, the allowed low-energy regions in the potential energy surface of each residue along the protein chain of length N. Within this mathematical construct, we may project the MD trajectory on the quotient space and then adopt deep learning to propagate the encoded "shorthand" dynamics to make computations more agile. The shorthand dynamics carry far lower information burden, while the transition timescale for each iteration is the intrabasin equilibration time $\tau=100\text{ps}$, considerably larger than the femtosecond integration timescales used in MD. Hence, since intrabasin events are lumped together and the iteration timescale is five orders of magnitude longer, the propagated modulo-basin dynamics has a far better chance of capturing the infrequent or rare interbasin transitions, a highly sought-after goal and a significant shortcoming in MD modeling of *in vitro* processes.

The deep learning process to propagate the modulo-basin dynamics uses a maximum likelihood (ML) ansatz to train a dilated convolutional network so that the ML weight, $f_\tau(n, b, b')$, for basin transition $b \rightarrow b'$ at chain position n after time τ elapsed, is influenced by the amino acid identity and basin occupancy at flanking residues $(n - 1, n + 1)$, with the influence field progressively dilated in the CNN layers so that flanking positions $(n - 2, n + 2)$, $(n - 3, n + 3)$,... , $(n - 8, n + 8)$ with respective strides $g = 1$, 2,..., 8 also impact the transition probability. Computational cost dictates that we limit the extent of dilation to maximum stride $g = 8$. Thus, the ML weight of the basin transition becomes:

$$f_\tau\left(n,b,b'\right) = \prod_{g=1}^{8} w\left(n,g,A\left(n,g\right),B\left(n,g\right)\right) f\left(n,g,A\left(n,g\right),B\left(n,g\right),b,b'\right),$$

(4.1)

where $f(n, g, A(n, g), B(n, g), b, b')$ gives the frequency of the $b \rightarrow b'$ basin transition at position n after time τ elapsed. This transition frequency is obtained by projecting MD runs onto a modulo-basin τ-discretized time series for chain composition $A(n, g)$ giving the amino acid identity of residues $n - g$, n, $n + g$ in the chain, and basin assignment $B(n, g)$ for residues $n - g$, n, $n + g$. The g-stride triads $(n - g, n, n + g)$ bearing on the basin transition at position n are weighted in accord with parameters $w(n, g, A(n, g), B(n, g))$, optimized relative to a suitable loss function. There are $20^3 \times 4^3 \times 8 = 4,096,000$ triad-weighting parameters to be optimized in the ML scheme.

The training of the network yields the transition frequencies $f(n, g, A(n, g), B(n, g), b, b')$, thus following the ML scheme. The propagation of the modulo-basin trajectory (time series) in the ML scheme is determined by the $b \rightarrow b'$ transition probabilities:

$$p_{ML}\left(n,b,b'\right) = \frac{f\left(n,b,b'\right)}{\sum_{b''} f\left(n,b,b''\right)},$$ where b'' is any of the four basins. The destiny

basin b' after a time period τ is determined by a Monte Carlo scheme taking into account the four transition probabilities at each position n along the chain.

The dilated CNN is thus organized into $8 \times 4^3 = 512$ layers in accord with stride g and basin assignment $B(n, g)$ for the triads centered at each position n on the chain and influencing the basin transition at position n. The input is a basin assignment for each residue $n = 1, ..., N$ along the chain, and the output is the basing assignment after time τ has elapsed. In turn, each layer contains $20^3 = 8000$ filters or convolution kernels in accord with the amino acid identity $A(n, g)$ of the triad. To convolve, a modulo-basin state for a chain of length N may be labeled with a 4N-vector $(b_1|b_2|...|b_n...)$ consisting of N binary 4-tuples b_n $(n = 1, 2, ..., N)$ indicating the Ramachandran basin occupancy for each residue, so value 1 at entry m in b_n $(b_{nm} = 1)$ indicates that residue n occupies basin m $(m = 1, 2, 3, 4)$. For example, $(1000|0010)$ indicates the coarse-grained state of a dipeptide with first residue in basin 1 and second residue in basin 3. Thus, a convolutional kernel $K(n, g, B_K(n, g))$ becomes a triad of binary 4-tuples (basin assignment $B_K(n, g)$) with stride g, with the convolution operation carrying the Kronecker-delta factor $\delta_{A(n, g), A_K(n, g)}$, where $A_K(n, g)$ is the kernel amino acid assignment for positions $n - g, n, n + g$. Each layer $L(q, g)$ is labeled by the basin triad $q = 1,..., 4^3 = 64$, and the stride g. In each layer, the $20^3 = 8000$ convolution kernels are slid along the chain as stencils, and full coincidence at position n in $L(q, g)$ yields concatenation (enrichment) at position n with the four attributes $f(n, g, A(n, g), B(n, g), b, b')$, one for each of the four a priori possible b' destiny basins. After further processing across all 512 layers, we arrive at a full representation of the chain at layer $L(64,8)$, where each position n is endowed with $8 \times 4 = 32$ transition parameters $f(n, g, A(n, g), B(n, g), b, b')$, one for each stride g and one possible destiny basin b'. In the 513th layer, the products $\Pi_{g=1}^8 w\left(n, g, A\left(n, g\right), B\left(n, g\right)\right) f\left(n, g, A\left(n, g\right), B\left(n, g\right), b, b'\right)$ are computed, and the four transition probabilities $p_{ML}(n, b, b')$ are obtained for each of the four a priori destinies b'.

Finally, in the output layer, the destiny basin for each residue is chosen using the Monte Carlo scheme.

The optimization of the field-influence factors $w(n, g, A(n, g), B(n, g))$ is carried out during the training phase of the modulo-basin propagator Γ_τ defined by the Monte Carlo generator of basin transitions. To train the network, we break up the MD trajectories spanning the time period $[0, t_f]$ into two regions, a training portion with timespan $[0, t_0]$ and an optimization portion covering the time interval $(t_0, t_f]$. With the trajectory statistics drawn for the region $[0, t_0]$, we obtain the basin-transition frequencies for triads with different stride centered at each particular residue. On the other hand, a variational optimization process during the period $(t_0, t_f]$ enables us to determine the w-parameters. Thus, the w-parameters are obtained from a commutativity condition that reflects the compatibility of fine and coarse-grained propagation of the trajectory and must be fulfilled during the period $(t_0, t_f]$. The commutativity condition is $\Gamma_\tau \circ \pi = \pi \circ M(\tau)$ as depicted by the following scheme:

$$x(t) \quad \xrightarrow{\pi} \quad \overline{x(t)}$$

$$\downarrow M(\tau) \qquad \downarrow \Gamma_\tau$$

$$x(t+\tau) \quad \xrightarrow{\pi} \quad \overline{x(t+\tau)}$$

Here, π is the canonical projection that assigns the Ramachandran basins to each backbone torsional state of the chain and $M(\tau)$ is the molecular dynamics propagator operating on torsional states of the chain. In other words, the parametrization of the modulo-basin propagator Γ_τ is optimized to minimize the loss function $\mathcal{L}(\Gamma_\tau)$ given by

$$\mathcal{L}(\Gamma_\tau) = M^{-1} \sum_{q=1}^{M} \| \pi x(t_0 + q\tau) - [\Gamma_\tau]^q \pi x(t_0) \|^2, \qquad (4.2)$$

where $M\tau \le t_f - t_0 < (M+1)\tau$. We know this loss function is the correct one since $\mathcal{L}(\Gamma_\tau) = 0$ if and only if the transition operator makes the diagram commutative. Once the optimal $\Gamma_\tau^* = argmin\ \mathcal{L}(\Gamma_\tau)$ has been obtained from stochastic steepest descent, the modulo-basin projected trajectory can be propagated beyond MD-accessible timescales. Physically meaningful timescales may be reached by computing the coarse states $x(t_f + q\tau)$ beyond the simulation timespan t_f as $[\Gamma_\tau^*]^q x(t_f)$.

The results of the AI-empowered propagation of the MD simulation of the structural adaptation of JNK within the WBZ_4/JNK association process are described in Figure 4.6. The initial state of the AI-empowered MD run incorporates the PDB-reported JNK structure where the dehydron M111-N114 is not formed, within a metastable complex obtained by positioning WBZ_4 relative to JNK according to the structural alignment with the cKIT/imatinib complex in PDB entry 1T46. The alignment is possible because kinases cKIT and JNK are homologous insofar as they belong to the same superfamily, and imatinib is the parental compound from which WBZ_4 is derived. The choice of initial condition is an educated guess justified since the binding mode of imatinib to kinase cKIT is expected to be similar to that of WBZ_4 with regards to JNK.

The AI-empowered MD simulation clearly reveals that WBZ_4 binding induces the dehydron M111-N114 (Figure 4.6) by contributing exogenously to its wrapping with the extra methyl not present in imatinib (Figure 4.5). Thus, we have once more corroborated the AI-based prediction in regards to the conformation-selecting modification of the imatinib scaffold. Not only *has AI taught the drug designer how to modify imatinib to target JNK but has also revealed the specific conformation of JNK induced upon drug/target association and the induced folding process that leads WBZ_4 to select the predicted conformation.*

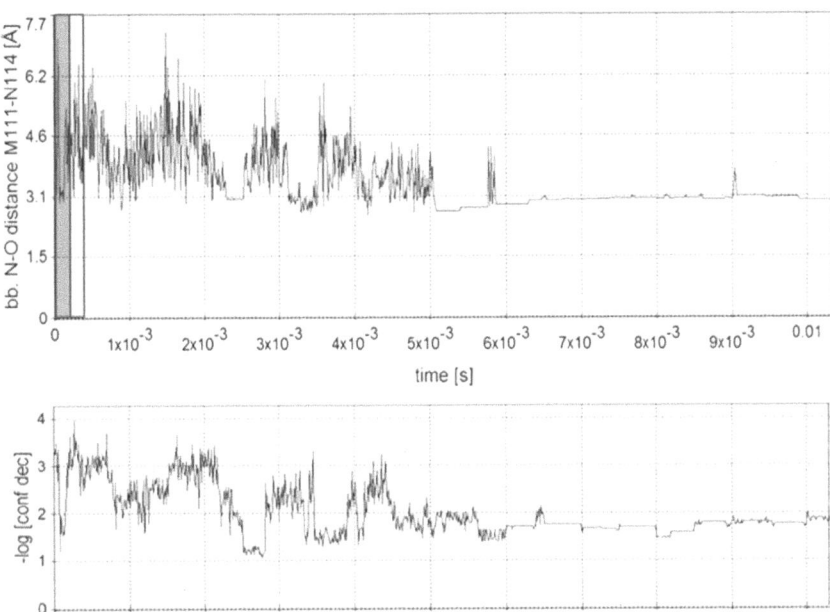

FIGURE 4.6 AI-empowered MD trajectory of the drug-induced folding trajectory for JNK in complex with WBZ_4 (Figure 12.3). The backbone N-O distance for residues M111, N114 in JNK is monitored along the trajectory which is given in the lower panel with the modulo-basin chain conformation represented in decimal form, as described in [14]. The BHB M111-N114 is clearly induced at about 6ms, thus signaling the cooperative binding where WBZ_4 methyl contributes exogenously to the wrapping of the BHB (Figure 12.3). The training-unassisted MD trajectory propagated with AI assistance is highlighted by the shaded box with its testing period in the unfilled box.

4.7 AI TEACHES DRUG DESIGNERS HOW TO TARGET PROTEINS BY EXPLOITING A DYNAMIC METAMODEL OF INDUCED FOLDING

Rational drug design faces difficulties, and it is felt it may not be living up to the expectations. This is mostly because the target proteins are usually not fixed targets: they structurally adapt to the ligand in ways that have been very hard to predict. That brings us to the drug-induced folding problem that is squarely addressed in this chapter aimed at fueling pharmaceutical discovery within an AI-empowered platform.

We implemented a deep learning platform to teach drugs to target proteins and identify the selected conformation of the target within the complex with the drug. The DL system is geared at steering drug design to induce targetable binding-competent protein conformations. The control of induced folding by a drug or ligand requires a DL platform that, at a variance with protein folding predictors like AlphaFold, integrates signals for structural disorder to predict conformations of floppy regions that rely on purposely designed binding partnerships to maintain their structural integrity. The conformational diversity of such regions begets an induced folding ensemble from which the targetable conformation gets selected by DL-instructed drug design, enabling a therapeutically significant drug-target association.

REFERENCES

1. Senior A W, Evans R, Jumper J, Kirkpatrick J, Sifre L, Green T, Qin C, Žídek A, Nelson AW, Bridgland A, Penedones H, Petersen S, Simonyan K, Crossan S, Kohli P, Jones DT, Silver D, Kavukcuoglu K, Hassabis D (2020) Improved Protein Structure Prediction Using Potentials from Deep Learning. *Nature* 577:706–710.
2. Dill KA, MacCallum JL (2012) The Protein Folding Problem, 50 Years On. *Science* 338:1042–1046.
3. Jones DT, Kandathil SM (2018) High Precision in Protein Contact Prediction Using Fully Convolutional Neural Networks and Minimal Sequence Feature. *Bioinformatics* 34:3308–3315.
4. Jones DT, Singh T, Kosciolek T, Tetchner S (2015) MetaPSICOV: Combining Coevolution Methods for Accurate Prediction of Contacts and Long Range Hydrogen Bonding in Proteins. *Bioinformatics* 31:999–1006.
5. Yu F, Koltun V (2016) Multi-Scale Context Aggregation by Dilated Convolutions. *arXiv:1511.07122*.
6. Chen JP, Liang H, Fernández A (2008) Protein Structure Protection Commits Gene Expression Patterns. *Genome Biol* 9:R107.
7. Pietrosemoli N, Crespo A, Fernández A (2007) Dehydration Propensity of Order–Disorder Intermediate Regions in Soluble Proteins. *J Proteome Res* 6:3519–3526.

8. Ovchinnikov V, Louveau JE, Barton JP, Karplus M, Chakraborty AK (2018) Role of Framework Mutations and Antibody Flexibility in the Evolution of Broadly Neutralizing Antibodies. *eLife* 7:e33038.

9. Fernández A, Bazan S, Chen J (2009) Taming the Induced Folding of Drug-Targeted Kinases. *Trends Pharm Sci* 30:66–71.

10. Schopf FH, Biebl MM, Buchner J (2017) The HSP90 Chaperone Machinery. *Nature Rev Mol Cell Biol* 18:345–360.

11. Plaxco KW, Simons KT, Baker D (1998) Contact Order, Transition State Placement and the Refolding Rates of Single Domain Proteins. *J Mol Biol* 277:985–994.

12. Fernández A, Scheraga HA (2013) Insufficiently Dehydrated Hydrogen Bonds as Determinants of Protein Interactions. *Proc Natl Acad Sci USA* 100:113–118.

13. Fernández A (2016) *Physics at the biomolecular interface: Fundamentals for molecular targeted therapy*. Springer International Publishing, Switzerland.

14. Fernández A (2020) Artificial Intelligence Steering Molecular Therapy in the Absence of Information on Target Structure and Regulation. *J Chem Inf Model* 60:460–466.

15. Obradovic Z, Peng K, Vucetic S, Radivojac P, Brown CJ, Dunker AK (2003) Predicting Intrinsic Disorder from Amino Acid Sequence. *Proteins: Struct Funct Gen* 53:566–572.

16. Oldfield CJ, Dunker AK (2014) Intrinsically Disordered Proteins and Intrinsically Disordered Protein Regions. *Annu Rev Biochem* 83:553–584.

17. Rampasek L, Goldenberg A (2016) TensorFlow: Biology's Gateway to Deep Learning. *Cell Sys* 2:12–14.

18. Druker BJ, Guilhot F, O'Brien SG, Gathmann I, Kantarjian H, Gattermann N, Deininger MW, Silver RT, Goldman JM, Stone RM, Cervantes F, Hochhaus A, Powell BL, Gabrilove JL, Rousselot P, Reiffers J, Cornelissen JJ, Hughes T, Agis H, Fischer T, Verhoef G, Shepherd J, Saglio G, Gratwohl A, Nielsen JL, Radich JP, Simonsson B, Taylor K, Baccarani M, So C, Letvak L, Larson RA, IRIS Investigators. (2006) Five-Year Follow-Up of Patients Receiving Imatinib for Chronic Myeloid Leukemia. *New Engl J Med* 355:2408–2417.

19. Fernández A, Scott LR (2017) Advanced Modeling Reconciles Counter-intuitive Decisions in Lead Optimization. *Trends Biotech* 35:490–497.

20. Kerkelä R, Grazette L, Yacobi R, Iliescu C, Patten R, Beahm C, Walters B, Shevtsov S, Pesant S, Clubb FJ, Rosenzweig A, Salomon RN, Van Etten RA, Alroy J, Durand JB, Force T (2006) Cardiotoxicity of the Cancer Therapeutic Agent Imatinib Mesylate. *Nature Med* 12:908–916.

21. Vivas-Mejia P, Benito JM, Fernández A, Han HD, Mangala L, Rodriguez-Aguayo C, Chavez-Reyes A, Lin YG, Carey MS, Nick AM, Stone RL, Kim HS, Claret FX, Bornmann W, Hennessy BT, Sanguino A, Peng Z, Sood AK, Lopez-Berestein G (2010) c-Jun-NH2-Kinase-1 Inhibition Leads to Antitumor Activity in Ovarian Cancer. *Clin Cancer Res* 16:184–194.

## CHAPTER 5

Targeting Protein Structure in the Absence of Structure

Metamodels for Biomedical Applications

To a man with a hammer
everything looks like a nail.

– Mark Twain

5.1 THERAPEUTIC DISRUPTION OF DYSFUNCTIONAL PROTEIN COMPLEXES IN THE ABSENCE OF REPORTED STRUCTURE

Protein-protein interfaces (PPIs) determine supramolecular complexes that get recruited in core biological processes in the cell, including signaling cascades, regulation of enzyme activity, and control of molecular motors [1]. A deregulated or dysfunctional complex may be disrupted through a molecular targeted intervention, offering a therapeutic opportunity [1, 2]. For example, the association of the negative regulator murine double-minute 2 homolog (MDM2) to the tumor suppressor p53 constitutes an important target in cancer therapies aimed at disrupting the modulation of p53 [2]. Another important illustration from cancer immunotherapy consists of disrupting PD-1 receptor recognition of its natural ligand PD-L1, a checkpoint blockade that enables the T-cell adaptive response by curbing tumor-induced immunosuppression [2].

DOI: 10.1201/9781003333012-7

Despite such successes, therapeutic disruption of PPIs remains challenging, mostly because the structure and regulated modulation of the complex are usually unknown or unreported. We address this problem by developing a deep learning (DL) platform [3] trained on structural [4] and inferred epistructural [5] data to identify regions on the protein surface capable of promoting functionally regulated protein associations. Such regions are not merely binding hot spots [5] but are also endowed with enzymatic activity to enable PPI modulation via the acquisition of signals in a regulatory setting. Thus, the underlying convolutional network exploits sequence-derived 1D-features [6] as input data, learns to infer structural and epistructural information, and uses the latter to generate an output that identifies the region fulfilling the dual role of binding epitope and regulation-susceptible element. We justify subsequently the need to go beyond a structural representation of the 1D-input for the purpose of discovery to disrupt deregulated PPIs. As in DL platforms for structural inference [7–9], the input is first processed into a 2D matrix of evolutionarily coupled residue pairs [10]. However, our approach differs fundamentally from other predictors [7–9] that are not concerned with identifying regulatory elements governing PPI modulation.

In this chapter, the coevolved-pair representation determines structural constraints that enable further training based on structure-derived features representing regulatory elements. The discovery platform is ultimately validated against experimental data on a therapeutic complex disruptor designed by mimicking the inferred epitope for a large dysfunctional complex [11–13] with unreported structure and poorly understood biochemical modulation.

If a binding partner in a protein complex becomes dysfunctional, complex formation may lead to disease, and hence PPI disruption represents a therapeutic imperative [1, 2, 14, 15]. An illustration this chapter focuses on is provided by the association of the myosin motor with the myosin-binding protein C (MyBP-C), a central regulator of cardiac contractility [12, 13]. When deregulated, the association promotes heart failure, while in healthy cardiomyocytes the PPI is modulated by phosphorylation at the MyBP-C binding epitope [12]. Thus, MyBP-C may be regarded as a "molecular brake" of the myosin motor. In the context of heart failure, a targeted intervention to release the brake constitutes an unmet therapeutic imperative [13] (Figure 5.1(a)). The intervention is tantamount to the drug-based disruption of a PPI, itself a major challenge [1, 2]. The design problem becomes daunting because the 3D structure of the multidomain MyBP-C

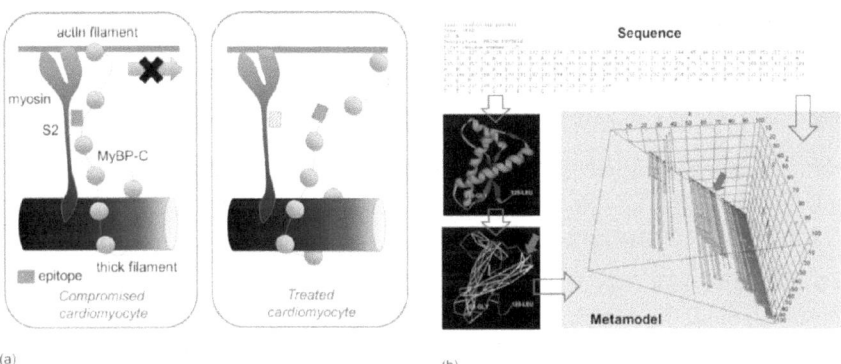

(a)   (b)

FIGURE 5.1   *Schematics of AI-guided therapeutic intervention to treat heart failure through targeted disruption of the PDB-unreported MyBP-C/myosin complex.* (a) The drug lead consists of a peptide reproducing the epitope region (shaded box) in MyBP-C. As the peptide binds to myosin competitively, it displaces the tethering/anchoring modulation inducing a "brake release" that accelerates the cross-bridge kinetics of cardiomyocyte contractility. Reprinted from [Fernández A (2020) Artificial Intelligence Steering Molecular Therapy in the Absence of Information on Target Structure and Regulation. Journal of Chemical Information and Modeling (ACS) 60:460–466] with permission from the American Chemical Society. (b) AI-based structural metamodel and its relation to the structure-based wrapping/dehydron model of a protein. The prion protein (PDB.1B10) serves as illustration. The structure-based model of the dehydron pattern is computed directly from the reported structure (appendix, section A.1), while the metamodel is obtained as a sequence-based inference made by an AI system that inputs the sequence-based dehydron inferences made by the *Twilighter* routine (Appendix, section A.2). Both the model and the metamodel identify the same dehydron-rich region (solid arrows) which constitutes a targetable epitope. This example corroborates the power of the AI-based methodology to steer molecular targeted therapy in the absence of structure for the target protein.

and the detailed regulatory context for the complex-modulation mechanism are unknown; hence, key issues like inter-domain binding cooperativity and induced folding [1] have remained essentially intractable.

Our goal is to steer molecular targeting by leveraging artificial intelligence (AI) technology [16, 17] trained on structural data to infer leads for therapeutic disruption of large dysfunctional complexes with unreported structure and regulation. In the specific context of heart failure, the goal is to identify a fragment of MyBP-C representing a peptide-based lead [15] that mimics a regulation-susceptible binding epitope. The peptide should be suitable for emergency interventions, administered in situ to displace MyBP-C from its association with myosin ([13], Figure 5.1(a)).

MyBP-C is a multidomain single polypeptide chain with a molecular weight of 140 kDa. Notwithstanding the amazing strides of AlphaFold in solving the protein folding problem, the hopes that DL technology could make a meaningful structure prediction for such a large multidomain protein are virtually nil at this point in time. Therefore, the kind of structure-based models of the wrapping pattern (Appendix, Section A.1) that can be used to identify the adhesive and hence targetable dehydron-rich regions are completely off limits for the current technology as applied to MyBP-C. What is needed in order to disrupt the aberrant complex is a way of inferring dehydron-rich regions in MyBP-C without resorting to a structure prediction. Since dehydrons belong to the twilight between order and disorder (Chapter 4), such inferences may be made with a bioinformatics tool named *Twilighter* that inputs the results of a sequence-based predictor of disorder (Appendix, section A.2). Thus, Twilighter predictions can be fed into an AI system that guides molecular therapy in the absence of structure by identifying a structural metamodel, that is, the wrapping tetrahedral tensor of the protein (Figure 5.1(b)). The example worked out in detail in this chapter illustrates the power of AI as a guide to molecular therapy.

To identify phosphorylation-susceptible binding epitopes from sequence-derived 1D-input, it is necessary to search beyond a structural description of the type that DL methods can achieve [7, 8], and represent the structure-solvent interface, that is the epistructure. To perform this task, we take advantage of the fact that the epistructural features are encoded in the structure itself [18–21], more specifically in the packing of the structure, that is, in the way backbone hydrogen bonds are shielded from structure-disruptive hydration [11]. The need to explore the epistructure is immediately apparent if we search for a binding hot spot since a binding region promotes the exclusion of interfacial water; hence, it generates interfacial tension [5]. On the other hand, identifying a phosphorylation-susceptible region also requires an epistructural representation, since the region must be endowed with functionalized side-chain hydroxyl groups, capable of launching a nucleophilic attack on the beta-gamma phosphodiester linkage in ATP [21]. So, the operative questions addressed in this chapter is: What epistructural feature may be a hot spot of interfacial tension while also promoting a nucleophilic attack on ATP, hence becoming a phosphorylation-based regulatory element? And how can we infer this epistructural feature in the absence of structure or information on regulation?

As it is known, a solvent-exposed backbone hydrogen bond (BHB), known as *dehydron*, fulfills both requirements [20, 21]. Water interfacing with a dehydron is deprived of full hydrogen-bond coordination in a manner that makes it also a proton acceptor, with hydronium becoming a good leaving group [18, 21]. Thus, due to suppressed hydrogen bonding of interfacial water, a dehydron creates interfacial tension, while it also functionalizes nearby side-chain hydroxyls (e.g., from serine, threonine, and tyrosine) by enhancing the basicity of coordination-deprived interfacial water to the point of accepting the hydroxyl proton [21]. As discussed subsequently, this dual role of dehydrons motivates the construction of a neural network sustaining a DL platform that provides an epistructural, rather than structural, output.

5.2 AI GUIDES THE THERAPEUTIC DISRUPTION OF A DYSFUNCTIONAL COMPLEX WITH NO REPORTED STRUCTURE

The discovery platform exploits a neural network trained on evolutionary and PDB-derived structural data [4] to represent the sequence-based 1D-input [6, 7] across a hierarchy of network layers up to an epistructural [5, 18] resolution that enables the identification of regulation-susceptible binding epitopes. Drug leads may be inferred from target sequence, provided DL is instructed on the epistructural characterization of the binding epitope. To that effect, we note that the MyBP-C-myosin complex is naturally disrupted upon phosphorylation of the so-called m-motif of MyBP-C, probably at loci Ser273, Ser282, Ser302 [12]. Thus, we focus on mimicking epitopes that promote water exclusion, driving protein associations, while also functionalizing spatially adjacent nucleophilic side chains of phosphorylation-susceptible residues as the epitope is exposed to water [21]. As described previously, dehydrons are precisely the type of epistructural features endowed with the sought-after dual role [18, 20].

To summarize, the overarching aim of this chapter is the development of DL method to expand the pharmaceutical discovery platform to cases where there is no detailed information on target structure and regulation of target activity. We focus on the most challenging cases involving rational design of molecular disruptors of deregulated protein complexes whose recruitment leads to disease. The neural network uses sequence-based 1D-input, gets trained on evolutionary and structural PDB-derived data and informed on how to generate epistructural features from structural inferences, and is capable of inferring binding regions that are also susceptible of phosphorylation-based regulation of their binding activity. This is possible because epistructural information or, equivalently, the structure-solvent dynamic relationship, which is key to infer regulated binding epitopes, is encoded in structural information. In turn, because of structural training, the network is able to represent the required epistructural information. Preliminarily, the deep learning discovery platform is benchmarked by establishing the accuracy of epistructural inference on a PDB-derived testing set as network connection parameters are optimized to minimize the appropriate loss function. Then, the DL platform is validated by showing it enables the identification of the clinically competent molecular disruptor of the deregulated MyBP-C-myosin complex recruited in the heart failure condition [13]. This peptide-based pharmaceutical has been tested *ex-vivo* and *in vitro* but the discovery process that led to its identification was not revealed in the patent that covered the invention.

5.3 REGULATORY SITES IN THE EPISTRUCTURE OF A PROTEIN TARGET

Since BHBs are structural determinants, protein integrity is contingent on the ability of the structure, by itself or within a complex, to prevent disruptive BHB hydration. Thus, dehydrons compromise structural integrity and promote protein associations to exclude surrounding water. Dehydrons become determinants for protein association because, by decreasing charge screening, exogenous removal of water from the dehydron microenvironment strengthens and stabilizes the electrostatic interaction that underlies the BHB. Ample bioinformatics evidence on the distribution of dehydrons at interfaces of protein complexes supports this picture, pointing to dehydrons as key factors driving complex formation. Dehydrons and epistructures are computed from structural coordinates of proteins. To that end, we introduce the extent of hydrogen-bond wrapping, w, indicating the number of side-chain nonpolar groups contained within a "desolvation domain" that defines the BHB microenvironment. Being marginally wrapped, dehydrons are located in the tail of the distribution of wrapping values across BHBs in the structural database.

In their alternative role, dehydrons play an enzymatic role by promoting phosphorylation of adjacent residues. This chemistry is enabled since dehydrons turn interfacial water molecules into proton acceptors, thereby activating the nucleophilic hydroxyls in the side chains of phosphorylation-susceptible residues (Figure 5.2). As the hydroxyl gets deprotonated, it is enabled to nucleophilically attack the β-γ phosphoester linkage of ATP, the key step in side-chain phosphorylation.

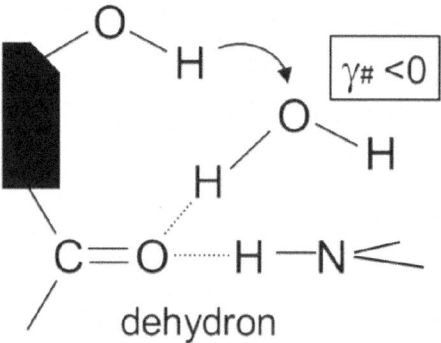

FIGURE 5.2   An epitope dehydron functionalizes a nearby side-chain hydroxyl by frustrating interfacial water, turning it into acceptor of the side-chain hydroxyl proton. Thus, dehydrons at epitopes can functionalize phosphorylation sites nearby, as required to modulate the MyBP-C/myosin interface.

To explain how dehydrons enhance basicity of nearby water requires a conceptual departure from the standard Debye picture where water polarization is assumed to align with electrostatic field determined by fixed charges [21]. The "Debye ansatz" breaks down under confinement of water molecules in regions of nanoscale dimensions, such as partially hydrated backbone regions at dehydron sites. The breakdown takes place since the bulk-like floppy lattice of water hydrogen bonds gets distorted to the point where water is effectively deprived of hydrogen-bond partnerships. Thus, dehydron-interfacing water molecules generate a polarization component $\overrightarrow{P^{\#}}$ orthogonal to the electrostatic field $\overrightarrow{E}$ and arising from partially hindered BHB hydration. This polarization induces an $\overrightarrow{E}$-independent net charge $\gamma^{\#} = -\overrightarrow{\nabla}.\overrightarrow{P^{\#}}$ not accountable by the protein chemical composition. The local dehydron-induced functionality of interfacial water is in accord with the sign of $\gamma^{\#}$, that is, proton donor if $\gamma^{\#} > 0$ and proton acceptor if $\gamma^{\#} < 0$. It has been shown that the latter possibility $\gamma^{\#} < 0$, implying that at least one water oxygen atom remains on average unpaired to a hydroxyl proton (its orbital-localized lone electron pair is unutilized), is the one that holds at dehydron sites (Figure 5.2). Thus, dehydron-interfacing water yields a non-Debye polarization-induced charge $\gamma^{\#} < 0$ that begets a proton-acceptor role compatible with unfulfilled hydrogen-bond coordination of water oxygen (where partial negative charge is located), turning dehydron-interfacing water into a chemical base. This result identifies dehydrons as promoters phosphorylation (Figure 5.2).

5.4 AI-BASED METAMODEL TO INFER REGULATION-MODULATED EPITOPES IN THE ABSENCE OF TARGET STRUCTURE

Leveraging a DL platform to identify dehydron-rich regions poses constraints on the represented data that may be processed by the network. The structural/epistructural data must lend itself to a tensor representation wherein information is describable and processed on a multidimensional array (tensor) of digital entries within a layered hierarchical architecture.

As argued subsequently, the DL algorithm queries the PDB-derived repository on predetermined evolutionarily coupled residues in order to represent BHBs involving the residue pairs. The training data has structural and epistructural components arranged in a tensor. The structural components are BHBs arranged in matrix form, while the epistructural components describe the extent of BHB wrapping. Epistructural components are thus drawn upon the structural data in tensor arrangement. The implemented learning algorithm pivots on two facts: (a) dehydrons constitute hot spots for phosphorylation-modulated protein association and (b) dehydron patterns lend themselves to a tensor representation whereby a 2D array (matrix) Y is constructed to indicate pairs of residues interacting through BHBs, while their respective wrapping r is represented along Y-orthogonal vectors, z_{ij} for entry Y_{ij}. Thus, the data processing realizes a tensor flow, with the tensor identified by the product $Y \otimes Z = [Y_{ij} \otimes z_{ij}]_{ij}$.

The DL flowchart is [sequence-derived 1D input features] ➔ [E = evolutionary coupling matrix compounded with 1D-features for paired residues] ➔ [structure/epistructure tensor $Y \otimes Z$] ➔ [scalar output identifying dehydron-rich region]. The DL network is trained to learn the relationship between evolutionarily coupled residues and dehydron-paired residues, with the goal of inferring dehydron-rich regions. The output gets further translated into a peptide-based drug lead that can be exploited for therapeutic disruption of specific protein associations, as illustrated in this study by the molecular targeted treatment of heart failure through disruption of the molecular brakes of cardiac contractility (Figure 5.1). Thus, the peptide-based drug lead embodies the inferred dehydron-rich epitope and enables therapeutic disruption of the dysfunctional complex [13].

The DL discovery platform is trained by the PDB-derived ASTRAL data set [4] filtered for sequences with less than 40% identity and parsed according to a 9:1 training:testing ratio, with no superfamilies intersecting both subsets. Network input nodes ..., x_i,..., x_j,... contain residue amino acid identity compounded with sequence-derived 1D-features including predicted solvent accessibility and predicted secondary structure [6]. Thus, the 1D-feature nodes are constructed by incorporating two dominant input features of the contact predictor MetaPSICOV, in turn obtained from predictors SOLVPRED (solvent exposure) and PSIPRED (secondary structure probability) [6]. For weak alignments, the contact predictor tends to overwrite the coevolution contribution in favor of the two 1D-features named above. Thus, for weak alignment, MetaPSICOV operates like a standard machine learning-based contact inference, whereas for strong alignments, coevolution signals outweigh the two standard 1D-features.

In the first layer (L_1), the information is transformed into a 2D array E representing the evolutionary coupling matrix (ECM) [7]. The rationale for using the evolutionary filter is that columns coupled in a multiple sequence alignment (MSA) will likely be close in the 3D structure. Pixel (entry) y_{ij} in the ECM informs on the extent of evolutionary correlation between residues in x_i, x_j, and this information is concatenated with the 1D-features in x_i, x_j. Once layer L_1 is constructed, the network is trained with the structural database. For each sequence/structure from the training data set, the E-matrix is computed along with the structure/epistructure information encoded in $Y \otimes Z$. With this training, the network learns to draw the relationship between evolutionary couplings y_{ij} and structural/epistructural features ($Y_{ij} \otimes z_{ij}$) (Figure 5.3). Thus, nodes in L_2 are elements in the symmetric tensor $Y \otimes Z = [Y_{ij} \otimes z_{ij}]_{ij}$, with "wrapping vector" $z_{ij} = [A_w(i, j)]_{w=1, 2, ...}$, where $A_w(i, j)$ denotes the number of learned representations of BHBs pairing residues in x_i, x_j with wrapping w (Figure 5.4(a)).

The output is a descriptor of the epistructure generated by the input sequence. It is defined as the set of scalars $<w>_i$, enabling inference of the regulation-susceptible binding epitope as the region richest in dehydrons with hydroxyl-containing side chains. The value $<w>_i$ indicates the expected extent of wrapping of BHBs involving the residue in x_i. The $<w>_i$ values are computed by compounding inputs drawn from layer L_2 resolved in the epistructural tensor (Figure 5.4(a)):

$$< w >_i = \left(\frac{1}{N-1} \right) \sum_{j \neq i, j=1}^{N} \sum_{w=1}^{M} w \frac{A_w(i, j)}{M(i, j)},$$

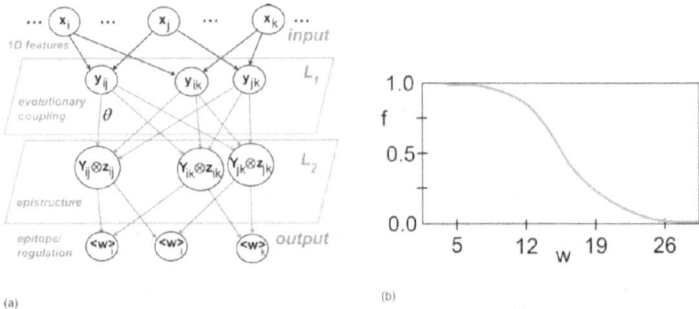

(a)   (b)

FIGURE 5.3 *Two-layer network for deep learning (DL) exploited in a drug discovery platform in the absence of structure for the target protein.* (a) DL network enabling the processing of input sequence information on the target protein and sustaining a tensor flow. In the first layer (L_1), the information is organized on the evolutionary correlation matrix $E = [y_{ij}]_{ij}$. The network is trained with structural and structure-derived epistructural information, yielding tensor $Y \otimes Z$ in layer L_2. By establishing correspondences between E and $Y \otimes Z$, the network learns the relationship between evolutionary coupling y_{ij} and backbone hydrogen-bond epistructure ($Y_{ij} \otimes z_{ij}$). The output is the set of scalars $<w>_i$, indicating the expected extent of wrapping of BHBs involving the residue in x_i. (b) Sigmoidal activation function (f) to represent the DL output. Reprinted from [Fernández A (2020) Artificial Intelligence Steering Molecular Therapy in the Absence of Information on Target Structure and Regulation. Journal of Chemical Information and Modeling (ACS) 60:460–466] with permission from the American Chemical Society.

where N is sequence length and $M(i,j) = \Sigma_{w=1}^{M} A_w(i,j)$, with M=32 being the maximum wrapping found in PDB. Thus, residue i is inferred to be paired by a dehydron if it belongs to the "wrapping-twilight region" $11.6 \leq <w>_i \leq 19.1$ (Figure 5.4(b)) [22].

5.5 STRUCTURAL METAMODEL FOR THERAPEUTIC DISRUPTION OF DYSFUNCTIONAL DEREGULATED PROTEIN COMPLEXES IN THE ABSENCE OF STRUCTURE

The DL algorithm is fully scripted in Python [23] with Tensor Flow [24]. The Python script is run from a PyMol 2.3 platform (Schrödinger, LLC, https://pymol.org/2/) as previously described (https://pymolwiki.org/index.php/Python_Integration), while PyMol exploits the dehydron calculator "Wrappy" as a plugin (Appendix 1). A two-layer network for DL is built and exploited in a drug discovery platform in the absence of structure for large multidomain target proteins. The DL network architecture

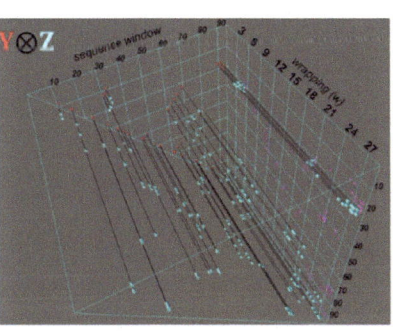

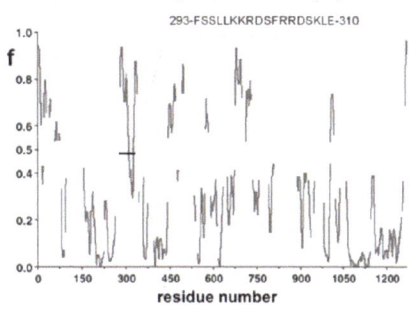

(a)

(b)

FIGURE 5.4   *Tensor representation of learned features and DL output for target MyBP-C sequence.* (a) Learned structural/epistructural $Y \otimes Z$ tensor for the testing 96-amino acid sequence in the initial Ig-domain (C0) of MyBP-C. The background matrix (Y) features the BHB-paired windows (Y_{ij}, red entries), while the learned wrapping vector z_{ij} (light blue entries) for each Y_{ij} is shown along the **Y**-orthogonal axis. The domain sequence processed is as follows: MPEPGKKPVS AFNKKP RSAE VTAGSA AVFE AETERSGV KV RWQRDGSDIT ANDKYGLA AE GKRHT LTVRD ASPDDQGSYA VIAGSSKVKF DLKVTE. (b) Deep learning output $<w>_i$ for x_i sliding along the amino acid sequence of MyBP-C represented with sigmoidal function f in Figure 5.3(b). There is inferred wrapping only for evolutionarily correlated residues learned to be paired by BHBs. The single region in the dehydron twilight $11.6 \leq w \leq 19.1$ is displayed. Reprinted from [Fernández A (2020) Artificial Intelligence Steering Molecular Therapy in the Absence of Information on Target Structure and Regulation. Journal of Chemical Information and Modeling (ACS) 60:460–466] with permission from the American Chemical Society.

enables the processing of input sequence information on the target protein and sustains the tensor flow [sequence-based 1D-input $\rightarrow$ **E** $\rightarrow$ $Y \otimes Z \rightarrow$ output]. In the first layer (L_1), the information is represented in an evolutionary coupling matrix $E=[y_{ij}]_{ij}$ using the tool CCMPred with default parameters [10]. Sequence alignments were generated with the tool HHBlits [25] using the UNIPROT sequence library [26], adopting e-value threshold at 0.001, pairwise identity cutoff at 0.99 and minimum coverage of the target sequence at 60%. The network is then trained using ASTRAL structure database [4] filtered for sequences with less than 40% identity and parsed according to a 9:1 training:testing ratio, with no superfamilies intersecting both subsets. Two sequence-derived 1D-features are compounded with amino acid identity and subsequently convoluted with evolutionary correlations for residue pairs in L_1. Given the desired output, the 1D-features chosen are predicted solvent accessibility computed by

SOLVPRED and predicted secondary structure, computed by PSIPRED, both run with default parameters [27].

Dehydrons and epistructures are identified from structural coordinates within the PyMol platform using the PyMol plugin [22] (Appendix 1). As described in Chapter 4, we define the extent of hydrogen-bond wrapping, w, indicating the number of side-chain nonpolar groups (CH_n, n = 1,2,3) contained within a "desolvation domain" that defines the BHB microenvironment. The desolvation domain is defined as two intersecting balls of fixed radius commensurate with three water layers centered at the α-carbons of the residues paired by the backbone hydrogen bond. A batch-mode wrapping analysis of the PDB revealed that 86% of backbone hydrogen bonds are wrapped by w = 26.6 ±7.5 nonpolar groups for a desolvation ball radius fixed at 6 angstrom. Dehydrons are located in the tails of the distribution, with their desolvation domain containing 19 or fewer nonpolar groups, so their w value is below the mean (w = 26.6) minus one standard deviation. More precisely, to ensure sustainability of the dehydron, the twilight region 11.6 ≤ w ≤ 19.1, corresponding to wrapping values in the range of one to two standard deviations below the mean.

By establishing correspondences between evolutionary couplings in **E** and structural/epistructural data in **Y** ⊗ **Z** for the training subset, the network learns the relationship between evolutionary coupling and BHB wrapping. The learning process is encoded in the connective weight optimization $\boldsymbol{\theta} = \boldsymbol{\theta}^* = argmin\ J(\boldsymbol{\theta})$, with loss function

$$J\left(\boldsymbol{\theta}\right) = \frac{1}{|\wp|}\sum_{s^{(n)} \in \wp} d\left[\boldsymbol{w}\left(\boldsymbol{s}^{(n)}\right), \boldsymbol{w}_{NN}\left(\boldsymbol{s}^{(n)}, \boldsymbol{\theta}\right)\right],$$

where $\wp$ denotes training set, $|\wp|$ is the number of elements in the set (cardinal), d indicates Euclidean distance, $\boldsymbol{s}^{(n)}$ is the nth sequence in $\wp$, $\boldsymbol{w}$ is the structure-derived wrapping vector and $\boldsymbol{w}_{NN}$ is the neural network wrapping inference. Thus, the minimization of $J(\boldsymbol{\theta})$ is a least-squares problem numerically solved with stochastic gradient descent. The set $\wp$ is sampled randomly at each iteration to compute the gradient with minibatch size $10^{-3} | \wp |$. To train the network, the ADADELTA method with learning rate 0.3 is adopted [28]. Full network training is shown in Figure 5.5, and takes approximately 16 hours on a cluster of 12 Titan RTX GPUs.

To validate and benchmark the DL discovery platform, native structures and structure-derived native wrapping patterns obtained from the ASTRAL-based testing set are contrasted against the neural network inferences generated from the respective primary sequences. To assess

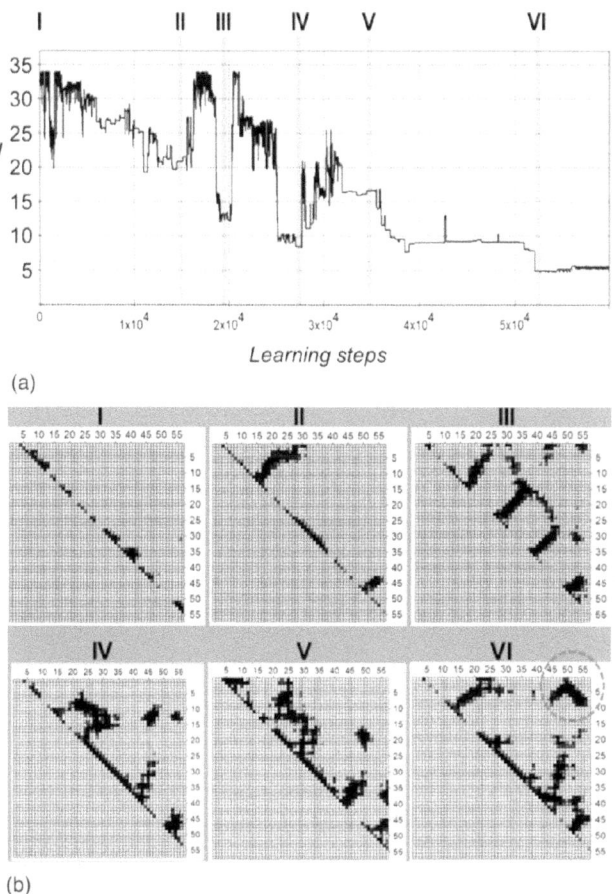

(a)

(b)

FIGURE 5.5 *Network training process.* (a) Loss function $J=J(\theta)$ with network connection weight vector θ optimized following stochastic gradient descent. Highlighted bands are regions of suboptimal (II-V) or optimal (VI) stability further examined by structural inference on the testing set adopting the respective parametrizations $\theta = \theta_m, m = I, II, \ldots, VI$. (b) Contact maps ($\beta$-carbon distance <8Å) for the hyperthermophilic variant of B1 domain in protein G (PDB.1GB4), a member of the testing set, learned from sequence-based evolutionary coupling with network parametrizations $\theta = \theta_m, m = I, II, \ldots, VI$. The emergence of the tertiary parallel β-sheet motif (circled) on panel VI is indicative of achieved accuracy, confirmed by comparing the topologies of the learned contact map in panel VI with

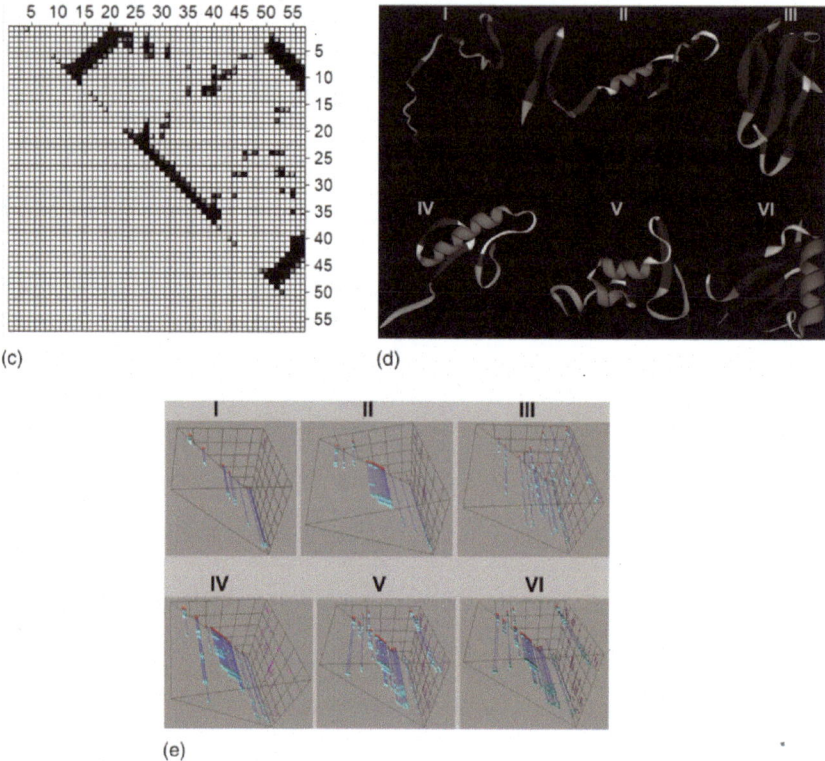

FIGURE 5.5 (*Continued*) (c), the native contact map for PDB.1GB4 (same folding topology). (d) Ribbon structural renderings of the learning process for test protein 1GB4. Structure in panel VI has backbone RMSD=1.54Å relative to the native fold in PDB.1GB4. The upper bound RMSD<2.45Å was found to hold across the entire testing set for parametrization $\theta = \theta_{VI}$, while native folding topologies were invariably reproduced, thus validating the machine learning method. (e) Learned epistructural tensor $Y \otimes Z$ for the six parametrizations along the learning process. Note that the learned wrapping level for each BHB is not unique, even for the optimized network, an indication of the sparsity of the training set vis-à-vis the dimensionality of the inferential space.

accuracy, a distance d between epistructural tensors $Y \otimes Z$ (PDB-derived) and $Y_{NN} \otimes Z_{NN}$ (inferred by the neural network) is introduced. Accordingly, d is defined as

$$d\left(Y \otimes Z, Y_{NN} \otimes Z_{NN}\right)$$
$$= \sum_{i<j}\left\{Q^{-1}\left[\Delta_{ij}\left(Y, Y_{NN}\right)\right] + q^{-1}\left[1 - \Delta_{ij}\left(Y, Y_{NN}\right)\right]\frac{\left|<w>_{NN,ij} - w_{ij}\right|}{M}\right\}$$

In the equation, Q is the number of BHBs that are either native PDB-reported or nonnative but inferred by the network; $\Delta_{ij}(Y, Y_{NN}) = 0$, if there are BHB pairing residues i and j in both native (Y) and inferred (Y_{NN}) structure, while $\Delta_{ij}(Y, Y_{NN}) = 1$ otherwise; $q = \Sigma_{i<j}\left[1 - \Delta_{ij}\left(Y, Y_{NN}\right)\right]$ is the number of correctly inferred native BHBs; M=32 is the maximum wrapping in PDB, as previously indicated; w_{ij} is the wrapping of the i-j BHB in the native structure; and $<w>_{NN,ij} = \Sigma_{w=1}^{M} w \dfrac{A_w(i,j)}{M(i,j)}$, $M(i,j) = \Sigma_{w=1}^{M} A_w(i,j)$,

is the expected wrapping for the i-j BHB inferred by the neural network (notation introduced previously). For each i,j-pair, the term $Q^{-1}\Delta_{ij}(Y, Y_{NN})$ contributes to the structural distance between native and inferred fold, while the second term $q^{-1}\left[1 - \Delta_{ij}\left(Y, Y_{NN}\right)\right]\dfrac{\left|<w>_{NN,ij} - w_{ij}\right|}{M}$ measures the wrapping distance between inferred and native structure for a correctly predicted i-j BHB ($\Delta_{ij}(Y, Y_{NN}) = 0$).

To benchmark our method, we note that the epistructural distance between native and inferred fold over the ASTRAL-derived testing set is invariably lower than 2.2% for an optimized parametrization of the network, with mean value at 1.4%. Of significance, at most 2% of native BHBs may be incorrectly predicted, while in cases where all native BHBs are correctly predicted, at most 1% of dehydrons are incorrectly predicted to be well-wrapped BHBs. These figures attest to the validity of the DL discovery platform.

The validation of the DL discovery platform is given in Figure 5.6.

5.6 EXPERIMENTALLY VALIDATING THE DYNAMIC METAMODEL OF THE TARGET PROTEIN STRUCTURE BY DEVELOPING A MOLECULAR TARGETED THERAPY FOR HEART FAILURE

The discovery platform is now validated by contrasting the epitope inference against a peptide-based lead adopted to treatment of heart failure [13]. In this therapeutic context, MyBP-C needed to be displaced from a large dysfunctional complex with myosin (Figure 5.1).

In healthy cardiomyocytes, MyBP-C reduces the speed and strength of contraction by interacting with the myosin S2-region, thereby decelerating cross-bridge kinetics by reducing the probability of myosin sliding along actin filaments [12, 13] (Figure 5.1). Phosphorylation of MyBP-C by kinases PKA or CAMKII disrupts this interaction and relieves the MyBP-C repressive tethering of myosin. On the other hand, in heart failure, MyBP-C is

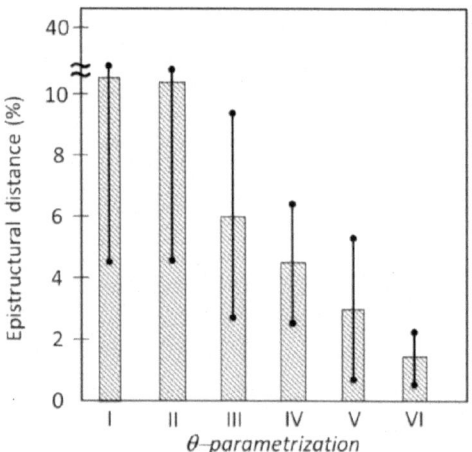

FIGURE 5.6 *Validation of the DL discovery platform.* Distance between network-inferred epistructure and native epistructure derived from PDB-reported structure from the testing subset. The inferences were made for six parametrizations along the optimization process to minimize the loss function, as shown in Figure 5.5. The bar indicates mean value, while the Gaussian dispersions are indicated by dumbbells. The DL neural network is trained by the PDB-derived ASTRAL data set filtered for sequences with less than 40% identity and parsed according to a 9:1 training:testing ratio, with no protein superfamilies intersecting both subsets.

phosphorylated minimally or not at all due to downregulation of β-adrenergic receptors. Based on the premise that disruption of the myosin-MyBP-C interface would release the molecular brake on cardiomyocyte contractility, we intend to displace MyBP-C with an optimal MyBP-C-derived peptide that materializes the DL-inferred epitope [13].

The lack of 3D structure precluded us from addressing design issues regarding the MyBP-C-myosin-S2 interaction, such as induced folding [18], inter-domain cooperativity and epitope-related functionalization of the phosphorylation sites in MyBP-C [21]. These limitations prompted us to discover the protein complex disruptor based on DL inference of the dehydron-rich region in MyBP-C that constitutes the putative S2-myosin-binding site. To predict the peptide sequence, a region was identified in the so-called m-motif between the MyBP-C domains C1 and C2 [12] fully contained in the dehydron twilight zone $11.6 \leq w \leq 19.1$ (Figure 5.4(b)). This is the only region inferred to be entirely in the wrapping twilight for MyBP-C. Thus, we identified the sequence of peptide ($_{293}$FSSLLKKRDSFRRDSKLF$_{310}$) that may be adopted, after S302A replacement to prevent phosphorylation, as emergency therapeutic agent to treat heart failure with administration in situ. The peptide shows significant increase in cardiac contractile force and frequency in the failing heart [13]. *Ex-vivo* experiments reveal that, upon peptide treatment, the contractile force in compromised cardiac tissue is restored to 82% of the force in normal tissue, compared with 44% in untreated compromised tissue. On the other hand, the rate of development of the contractile force is restored to 84% levels, compared with 22% for untreated compromised tissue [13].

We have highlighted the power of artificial intelligence (AI) to guide therapeutic interventions in clinical settings where a clear molecular understanding of the dysfunctionality is lacking.

Drug-based disruptions of PPIs in large dysfunctional complexes represent therapeutic opportunities but pose great challenges, as ligands are required to displace binding partners with extensive interfaces. The problem becomes daunting because the structure of the protein complex and its regulated modulation are frequently unknown and the complex often has insufficient homologs to build a reliable model of the PPI. To address the challenge, we developed a DL platform trained with big data on structural defects in proteins. These defects are known to cause interfacial tension [5, 18] and to constitute regulatory elements [21]. The network exploits sequence-based input and learns to draw a relationship between evolutionarily coupled residues, epistructural features, and regions that promote water exclusion through protein-protein association, ultimately becoming capable of inferring phosphorylation-modulated binding epitopes.

This chapter thus paves the way to an expansion of the technological base for rational drug design. The developments are particularly suited to therapeutic contexts where target structure and detailed information on regulation of target activity are missing or underreported.

The DL discovery platform implemented in this work enabled focusing on some of the most challenging cases, involving the identification of molecular disruptors of dysfunctional PPIs, whose recruitment is conducive to disease. The neural network uses sequence-derived 1D-input, gets trained on evolutionary and structural PDB data, and generates epistructural information that leads to the identification of binding regions susceptible of phosphorylation-based regulation. These regions generate

interfacial tension and functionalize nearby phosphorylation-susceptible side chains; thus, they become essential to create peptide-based PPI
disruptors within a rational discovery effort. The discovery is made possible because epistructural information or the structure-solvent dynamic
relationship is encoded in structural information, which the network is
able to infer from 1D sequence-based input. At a preliminary level, the
DL discovery platform is benchmarked by establishing the accuracy of
epistructural inference on a PDB-derived testing set. Subsequently, the
DL platform is validated through the identification of the clinically competent peptide-based disruptor of the deregulated MyBP-C-myosin
complex that gets recruited in heart failure. The therapeutic efficacy of
this peptide-based pharmaceutical has been previously established but
the discovery pathway that led to its identification was not revealed in
the patent where the invention is described [13].

5.7 THE DYNAMIC METAMODEL OF PROTEIN STRUCTURE ENABLES DISCOVERY OF BIOLOGICAL COOPERATIVITY

There are obviously many ways in which artificial intelligence (AI) is
likely to broaden the technological base of the pharmaceutical industry. Applications of AI that involve the parsing of the vast chemical
space are already being implemented. This book takes a complementary
view and advocates that the true game change comes with AI-steered
solutions to core problems in biophysics that are intimately related to
drug discovery and development. These problems share a common
challenge: understanding and controlling cooperativity, a hallmark
of biological processes and drug-target associations. A take-home les-
son from this book is that drug design guided by engineering pairwise
interactions across a putative drug-target interface is not likely to be
successful. Drug designers are actually faced with a many-body problem, where correlations are indicative of cooperativity and cooperative
patterns need to be exploited because they represent filters for drug
specificity while giving the complete more accurate picture of drug-
target complexation. Thus, the designers need to master this aspect
if their compounds are expected to withstand the long-term attrition
along the discovery pipeline. This closing informal section, tailored as
a short essay, argues that precisely because cooperativity can be cast
in tensor format, AI is likely to play a decisive role in unraveling and

predicting patterns of cooperativity at molecular scales, hence transforming, hopefully revolutionizing, pharma as we know it.

Broad examination of biomolecular architectures in aqueous media reveals a stunning complexity, where the structural determinants are hydrogen bonds (HBs) pairing identifiable fixed partnering groups. This observation has turned into a *leitmotif* in the organization of information in structural biology, particularly when dealing with protein folding, binding-induced folding, protein complexation, and protein-ligand binding. However, it has been argued that the potential energy function governing such interactions is not additive and thus cannot be accurately described as sums of pairwise contributions. The exclusion of water, partial or complete, from the surroundings of an HB is essential to warrant its stability and strength, and this fact alone signals that we are dealing with a many-body problem underlying a highly cooperative process: no pairwise interaction per *se* is likely to prevail unless it forms cooperatively so as to stay "dry in water." We may say that in this picture, protein folding, binding-induced folding, and complexation may be regarded as "struggles for the survival of HBs." Cast more rigorously, protein structure in water is only sustainable if backbone HBs are shielded from the disruptive hydration of amide and carbonyl groups. This implies that the structure must often rely on binding partnerships to maintain its integrity. This is the case, for example, with the binding-induced folding of transcription factors, which are autonomously incapable of folding into stable conformations. In other instances delineated in Chapter 4, the soluble structure is autonomously untenable due to backbone exposure and the protein misfolds into a conformation that can only be stabilized at high concentrations through aggregation, as it is the case in degenerative neuropathies with an aberrant protein aggregation ethiology.

We are clearly dealing with a many-body problem: The concurrence of additional groups is required to shield a pairwise electrostatic interaction from the bond-debilitating dielectric modulation and the disruptive effect of amide and carbonyl hydration. These patterns are in fact three-body correlations that bespeak of a cooperative process of structure formation: Secondary structure cannot form incrementally but needs either tertiary-structure buttressing or exogenous protection, as in protein complexes. This shielding from water is what we have named and quantified as *wrapping* (Chapters 3 and 4). Adopting a metaphoric language, the nonadditivity of the potential energy indicates that "the whole is actually more than the sum of the parts", while protein folding is reminiscent of Rubik's cube moves, where the solver does not work on one face at time but on all of them at once when deciding each move.

Let us illustrate this elusive aspect of Nature's architecture. An exhaustive cross-examination of the PDB and the sequence-based predicted propensity for structural disorder reveals that folding cooperativity is nowhere more apparent than in ubiquitin (PDB.1UBI). The reported structure features a cleanly formed helix spanning residues 22 to 35 (Figure 5.7). By contrast, predictors of native disorder (PONDR) indicate inherent disorder for the 22–35 region (Figure 5.8), clearly at odds with the structural data. This clash suggests a highly cooperative process that cannot be captured by the locality of the disorder prediction. The predictor slides a window along the protein sequence and draws accordingly the plot of disorder propensities (Figure 5.8). Indeed, a profusion of three-body correlations with distant nonhelical residues (Figures 5.9 and 5.10) reveals a highly cooperative shielding of backbone HBs in the helix region (22–35) that would be otherwise unsustainable in water.

As shown in this chapter, AI is instrumental in capturing such cooperative patterns, classifying them, and accordingly organizing structural information. Telling evidence based on analysis of the connectivity parameters for the dilated convolutional neural network (CNN) described

PDB.1UBI

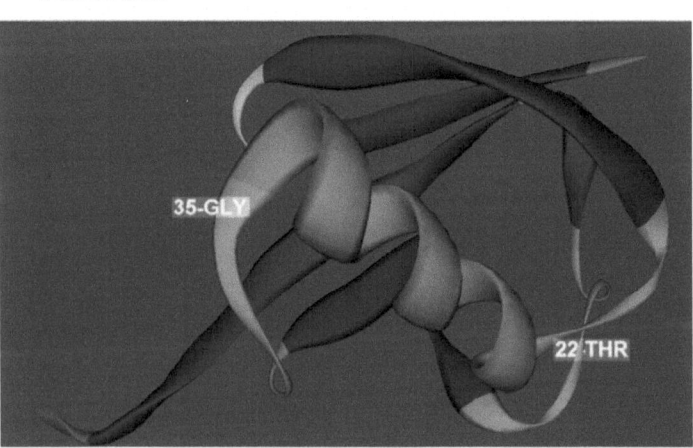

FIGURE 5.7   Ribbon rendering of the structure for human ubiquitin as reported in PDB entry 1UBI.

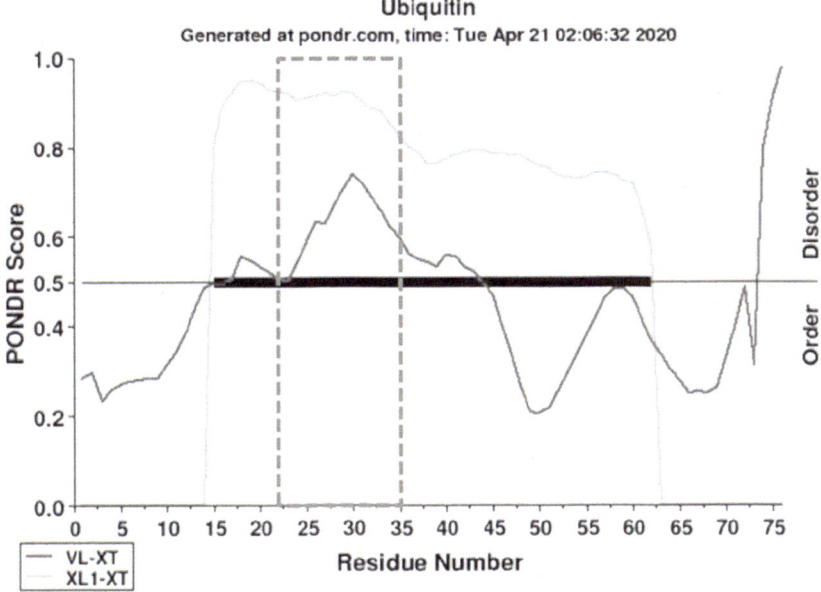

FIGURE 5.8 Two sequence-based predictions of intrinsic disorder for ubiquitin obtained from PONDR (http://www.pondr.com/). The dashed rectangle marks the sequence window 22–35, known to be helical but predicted to be disordered.

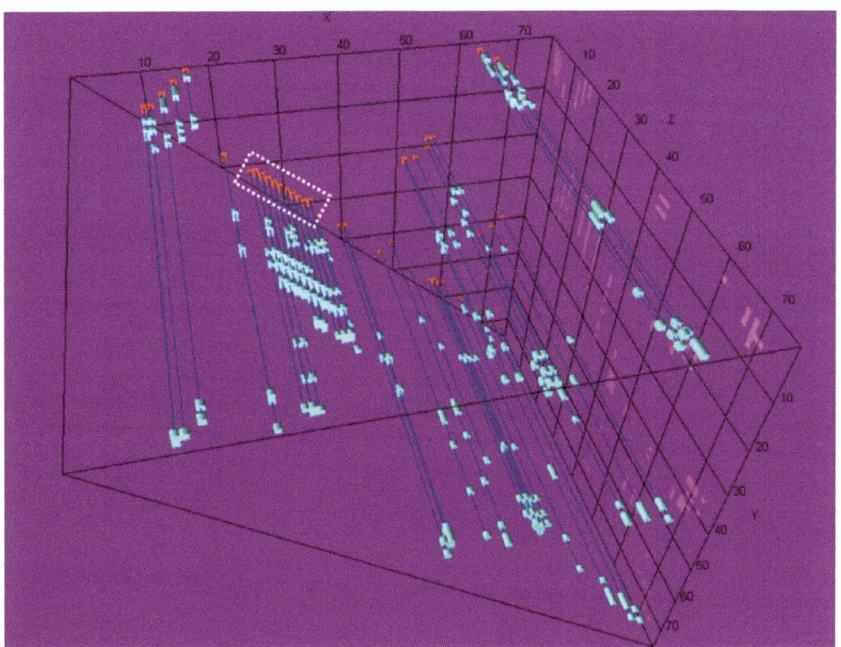

FIGURE 5.9 Wrapping tensor for ubiquitin structure. The dashed rectangle highlights the highly cooperative helical region 22–35.

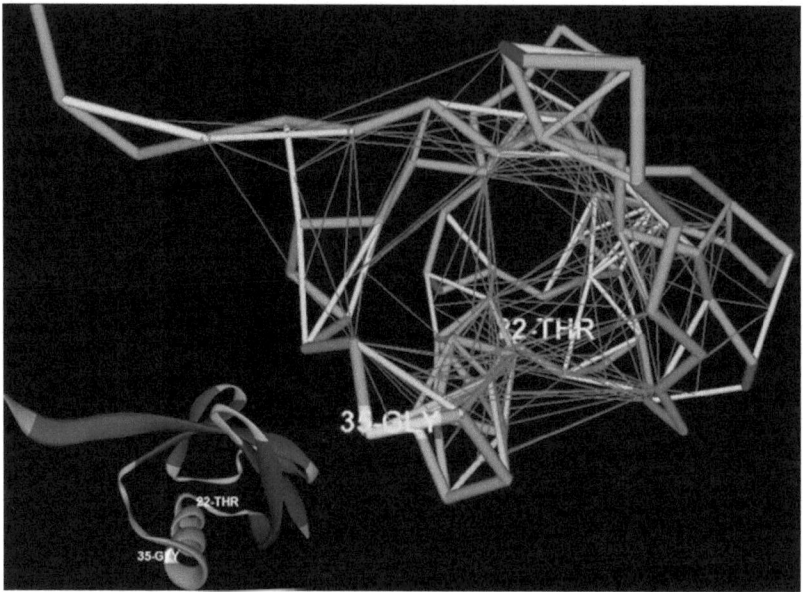

FIGURE 5.10 Wrapping pattern for helix 22–35. The profusion of three-body correlations involving distant nonhelical residues indicates that the helix forms in a highly cooperative way.

in this chapter, supports this assertion. In fact, all the three-body correlations involving backbone HBs in the ubiquitin 22–35 helix represented in the wrapping tensor (Figure 5.9) are actually contained in the optimized convolution kernels or filters adopted by the deep learning system for feature extraction in sequence-based inference of induced folding. To represent this assertion, three-body correlations were linked in the 3D wrapping tensor whenever they were contained in the same filter in the dilated CNN. Thus, as shown in Figure 5.11, the wrapping three-body correlations in the most highly cooperative helix in the PDB have all been inferred independently by AI, since they are all contained in the filters of the dilated CNN that have evolved to perform the optimal feature extraction.

This analysis is indicative of the power of AI to identify patterns of cooperativity, hence unraveling a major hallmark of biomolecular processes and drug-target associations. This instills confidence that the true game change in pharmaceutical discovery may be within arm's reach.

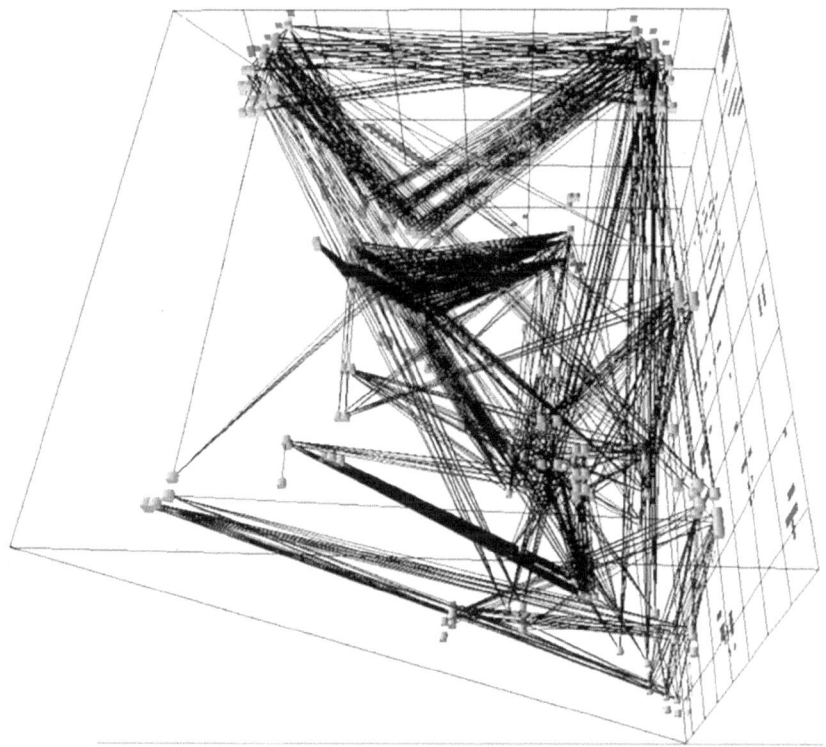

FIGURE 5.11 The convolution kernels in the dilated CNN for the deep learning system displayed on the ubiquitin wrapping tensor. Two three-body correlations are linked by a black line if they belong to the same optimized filter in the feature-extraction phase. Thus, the most highly cooperative patterns had been captured by the deep learning system.

REFERENCES

1. Scott DE, Bayly AR, Abell C, Skidmore J (2016) Small Molecules, Big Targets: Drug Discovery Faces the Protein-Protein Interaction Challenge. *Nat Rev Drug Discov* 15:533–550.
2. Li Z, Ivanov AA, Su R, Gonzalez-Pecchi V, Qi Q, et al. (2017) The OncoPPi Network of Cancer-Focused Protein–Protein Interactions to Inform Biological Insights and Therapeutic Strategies. *Nature Comm* 8:14356.
3. Sejnowski TJ (2018) *The deep learning revolution*. MIT Press, Cambridge
4. Fox NK, Brenner SE, Chandonia JM (2013) SCOPe: Structural Classification of Proteins – Extended, Integrating SCOP and ASTRAL Data and Classification of New Structures. *Nucleic Acids Res* 42:D304–D309.
5. Fernández A (2012) Epistructural Tension Promotes Protein Associations. *Phys Rev Lett* 108:188102.

6. Jones D, Singh T, Kosciolek T, Tetchner S (2015) MetaPSICOV: Combining Coevolution Methods for Accurate Prediction of Contacts and Long Range Hydrogen Bonding in Proteins. *Bioinformatics* 31:999–1006.

7. Gao M, Zhou H, Skolnick J (2019) DESTINI: A Deep Learning Approach to Contact-Driven Protein Structure Prediction. *Scientific Reports* 9:3514.

8. Wang S, Sun S, Li Z, Zhang R, Xu J (2017) Accurate de Novo Prediction of Protein Contact Map by Ultra-Deep Learning Model. *PLoS Comput Biol* 13:e1005324.

9. Liu Y, Palmedo P, Ye Q, Berger B, Peng J (2018) Enhancing Evolutionary Couplings with Deep Convolutional Neural Networks. *Cell Syst* 6:65–74.e63.

10. Seemayer S, Gruber M, Soding J (2014) CCMpred – Fast and Precise Prediction of Protein Residue-Residue Contacts from Correlated Mutations. *Bioinformatics* 30:3128–3130.

11. Fernández A, Scheraga HA (2013) Insufficiently Dehydrated Hydrogen Bonds as Determinants of Protein Interactions. *Proc Natl Acad Sci USA* 100:113–118.

12. Kensler RW, Craig R, Moss RL (2017) Phosphorylation of Cardiac Myosin Binding Protein C Releases Myosin Heads from the Surface of Cardiac Thick Filaments. *Proc Natl Acad Sci USA* 114:E1355–E1364.

13. Moss RL, Fernandez A (2015) Inhibition of MyBP-C binding to myosin as a treatment for heart failure. United States Patent 9051387.

14. Arkin MR, Tang Y, Wells JA (2014) Small-Molecule Inhibitors of Protein–Protein Interactions: Progressing Toward the Reality. *Chem Biol* 21:1102–1114.

15. Pelay-Gimeno M, Glas A, Koch O, Grossmann TN (2015) Structure-Based Design of Inhibitors of Protein–Protein Interactions: Mimicking Peptide Binding Epitopes. *Angew Chem Int Ed Engl* 54:8896–8927.

16. Wang J, Cao H, Zhang J, Qi Y (2018) Computational Protein Design with Deep Learning Neural Networks. *Scientific Reports* 8:6349.

17. Chen H, Engkvist O, Wang Y, Olivecrona M, Blaschke T (2018) The Rise of Deep Learning in Drug Discovery. *Drug Discovery Today* 23:1241–1250.

18. Fernández A (2016) *Physics at the biomolecular interface*; Springer, Berlin.

19. Fernández A, Rogale K, Scott LR, Scheraga HA (2004) Inhibitor Design by Wrapping Packing Defects in HIV-1 Proteins. *Proc Natl Acad Sci USA* 101:11640–11645.

20. Fernández A (2018) Stickiness of the Hydrogen Bond. *Ann Phys (Berlin)* 530:1800162.

21. Fernández A (2014) Communication: Chemical Functionality of Interfacial Water Enveloping Nanoscale Structural Defects in Proteins. *J Chem Phys* 140:221102.

22. Martin O (2014) Wrappy: A Dehydron Calculator Plugin for PyMOL. MIT License, 2014 http://www.pymolwiki.org/index.php/dehydron.

23. Ketkar N (2017) *Deep learning with python*. Apress, Berkeley, CA.

24. Rampasek L, Goldenberg A (2016) TensorFlow: Biology's Gateway to Deep Learning. *Cell Sys* 2:12–14.

25. Remmert M, Biegert A, Hauser A, Söding J (2011) HHblits: Lightning-Fast Iterative Protein Sequence Searching by HMM-HMM Alignment. *Nature Methods* 9:173–175.

26. The UniProt Consortium (2019) UniProt: A Worldwide Hub of Protein Knowledge. *Nucleic Acids Res* 47:D506–515.
27. Jones DT, Kandathil SM (2018) High Precision in Protein Contact Prediction Using Fully Convolutional Neural Networks and Minimal Sequence Feature. *Bioinformatics* 34:3308–3315.
28. Zeiler MD (2012) ADADELTA, an adaptive learning rate method. arXiv:1212.5701. https://arxiv.org/abs/1212.5701

## CHAPTER 6

Autoencoder as Quantum Metamodel of Gravity

Toward an AI-Based Cosmological Technology

The physical world is only made of information, energy and matter are incidentals.

– John A. Wheeler

6.1 THE QUEST FOR QUANTUM GRAVITY

A theory of quantum gravity (QG), that is, a theory of gravity in accord with the tenets of quantum mechanics (QM), is regarded as the holy grail by a certain breed of physicists [1, 2]. This is because all forces of nature, except gravity, can be understood through the quantum laws, and hence a sense of incompleteness pervades the field. Thus, a theory of QG portends to unify all the forces of nature.

A different sort of grand unification was pursued by Albert Einstein. He sought to do it without involving QM, a theory he never fully endorsed. At first glance, quantum gravity stands almost as an oxymoron: after all, QM deals with the atomic and subatomic scales, while the best theory of gravity to date is Einstein's general relativity (GR), which is essentially classic, that is, nonquantum, and deals mainly with cosmological scales (except for singularities). Einstein's theory of relativity postulates that high concentrations of energy and matter impinge on the curvature of space-time, deflecting the trajectories of particles, as it occurs in a gravitational field. This theory withstood admirably the long-term attrition of experimental corroboration and keeps being validated time after time. Yet, if we attempt to cast GR in QM terms, we not only stumble upon the scaling problem, but we need to deal with the fact that matter and space-time become "protean" at scales of the order of Planck's length (10^{-33} cm), akin to the sea of virtual particles that fill up empty space. In this quantum world, the equations of GR no longer hold valid, in fact they seem hopelessly inadequate. At first glance, the ethos of this book, emphasizing the need for physical models without equations seems to resonate with this inherent and daunting difficulty with QG.

Perhaps pursuing QG still makes good sense as we seek to treat black holes or the first few nanoseconds after the big bang, when the vast contextual differences between QM and GR can be reconciled. After all, these differences are in fact reconciled in the reality of such singularities in space-time. Thus, in the spirit of this book, we need to address the question: *What would constitute a quantum metamodel for a theoretical model of gravity and how to construct it?* The question posed in this way has been addressed to some extent by Argentinian physicist Juan Maldacena [1]. He focused on the so-called anti de Sitter (AdS) space, a hyperbolic space that shares curvature properties with the sphere representing the event horizon of a black hole. Maldacena postulated that a string theory (ST) of gravity in a five-dimensional AdS space (AdS_5 = W in standard notation) is *equivalent* to a quantum field theory (QFT) on its boundary ∂W, which constitutes a four-dimensional Minkowski space, akin to the one adopted by GR.

Why would the boundary represent the latent space for the metamodel of the ST in an AdS? The answer is provided by the computation of the Shannon entropy of the black hole that assesses its total information storage capacity enshrined in all degrees of freedom [3]. These "ultimate" degrees of freedom involve of course atomic and subatomic entities, all the way to quarks and gluons, and ultimately those entities from hitherto unknown depths in the structural description of matter, the so-called level X. It is well known that this "ultimate entropy" is proportional to the surface area of the event horizon.

In general, the boundary M_d of AdS_{d+1} is a d-dimensional Minkowski space with the symmetry group $SO(2, d)$ of AdS_{d+1} acting on M_d as the conformal (i.e., inner product-preserving) group. Thus, there are two ways to get a physical theory with $SO(2, d)$ symmetry: a relativistic field theory on AdS_{d+1} and a conformal field theory (CFT) on M_d. A suitable theory on AdS_{d+1} has been conjectured by Maldacena to be *equivalent* to a CFT on M_d [1]. The computation of observables of the CFT in terms of supergravity on AdS_{d+1} can and should be attempted using the methods described in this book. In accord with the tenets of topological metamodel autoencoding, M_d should be identified with the latent manifold Ω and the CFT on M_d "holographically" spanned onto AdS_{d+1} should be generated by a holographic autoencoder that exploits the $SO(2, d)$ to generate jointly the latent manifold together with its parsimonious metamodel.

As stated by Witten, correlation functions in quantum field theory are given by the dependence of the supergravity action on the asymptotic behavior at infinity [2]. Thus, dimensions of operators in CFT are determined by masses of particles in string theory. It is thus conjectured that to describe the Yang-Mills theory in four dimensions, one should use the whole infinite tower of massive Kaluza-Klein states on $AdS_5 \times \mathbf{S}^5$. Chiral fields in the four-dimensional $N = 4$ theory correspond to Kaluza-Klein harmonics on $AdS_5 \times \mathbf{S}^5$. The spectrum of Kaluza-Klein excitations of $AdS_5 \times \mathbf{S}^5$ are matched against operators of the $N = 4$ theory.

To discover the topological metamodel for superstring theory on $W = AdS_{d+1} \times \mathbf{S}^{d+1}$, we first note that the boundary is topologically identified as $\partial W = (S^1 \times S^{d-1})/\mathbb{Z}_2$, where the group $\mathbb{Z}_2$ acts by rotation in π on S^1 and multiplication by -1 on S^{d-1}. In other words, the latent manifold fulfills the compactness condition (Chapter 2), and to simplify the computation and straighten the symmetry we may use its universal cover [4]: $\Omega \approx S^{d-1} \times \mathbb{R}$, with the real axis representing the time dimension.

Thus, to prove using AI the conjectured equivalence between $N = 4$ QFT on $\Omega \approx S^{d-1} \times \mathbb{R} = \partial W$ and Type IIB supergravity as string theory (ST) on $W = <AdS_5 \times \mathbf{S}^5>$ (<.> = universal cover [4]), we need a holographic autoencoder for a DL neural network capable to represent the ST on the space W. This DL system has been constructed [5]. The autoencoder should identify the latent manifold as $\Omega \approx S^3 \times \mathbb{R}$ taking advantage of the $SO(2,4)$ symmetry with which the AdS space is endowed. In this way, the projection π onto the latent manifold, identified by the autoencoder as ∂W,

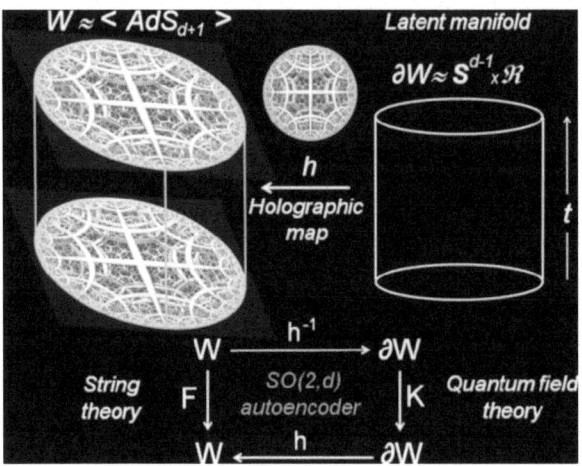

FIGURE 6.1 Holographic autoencoder enabling the discovery of a quantum metamodel of gravity.

constitutes the inverse of the postulated holographic map $h : \partial W \to W$ that makes the following diagram commutative (Figure 6.1):

$$
\begin{array}{ccc}
W & \xrightarrow{\pi} & \partial W \\[2ex]
\downarrow F_{ST} & & \downarrow K_{QFT} \\[2ex]
W & \xleftarrow{h} & \partial W
\end{array}
$$

Thus, the AdS/CFT equivalence may be proved by AI. The program entails two steps: (a) generation of a data-driven holographic metamodel of quantum systems by formulating its supergravity dual/equivalent on a DL NN and (b) construction of the appropriate autoencoder (π, K_{QFT}) that fits the holographic map h, yielding the functional identity $h \circ K_{QFT} \circ \pi = F_{ST}$ of homeomorphisms on W. Part (a) of the program has already been achieved, as shown elsewhere [5]. A deep NN representation of the AdS/CFT correspondence has been obtained, with emergence of the bulk metric based on deep learning of data generated as outcome of boundary quantum field theories. The radial direction of the bulk metric is assimilated with the depth of the hidden layers. Thus, the network provides a data-driven model of strongly coupled systems. In the space-time for a black hole horizon, the deep NN can fit boundary data generated by the AdS Schwarzschild space-time, reproducing the metric as the data is reverse engineered. With inputted experimental data, the deep NN determines the bulk metric, mass and quadratic coupling. Thus, the NN provides a gravitational model of strongly correlated systems.

As for part (b), the HAE that will ultimately identify the latent manifold with ∂W can be obtained using the methods described in Chapters 1 and 2. The remaining challenge in part (b) is for the HAE to simultaneously yield the parsimonious QFT at the boundary ∂W that serves as metamodel for the gravity model on W in the sense adopted in this book. To address this challenge requires that we harness the isomorphism already noted [6] between a deep autoencoder and the multiscale entanglement renormalization ansatz (MERA).

The research proposal described attests to the generality of the methods expounded in this book. If the universe is indeed a hologram, as the unified theory of all forces of nature seems to suggest, the inverse of the holographic map of its quintessential avatar, the event horizon of the black hole, represents a 1–1 (injective) projection onto the horizon boundary. Therefore, this boundary may be regarded as a model for the latent quotient manifold of the universe. In this quotient manifold, one dimension has been folded up and stored within the equivalence classes that are ultimately the only observables and hence the only encoded features endowed with physical entity.

Ultimately, quantum gravity will be ascertained through a holographic autoencoder that identifies the "correct" latent manifold and associated quantum metamodel, or perhaps by other means available to theoretical physicists. Be as it may, we may state that all events following the Planck epoch of the big bang ($t > 10^{-43}$ s [7]) are likely to be of a quantum nature. This is so because at that stage gravity branches off from the three other fundamental forces already accounted for by quantum theory, so all four differentiated forces will be reliably identified as quantum forces. Thus, the certainty of all events that follow the Planck epoch can only be established through the participation of observers. This singular circumstance leaves us having to postulate god's coming into existence at least as early as 10^{-43} s after the big bang, or admitting that the universe remains a mere possibility, replete with phenomena-to-be within multiple a priori realities that are equally possible in the quantum realm, as in a multiverse scenario.

6.2 QUANTUM GRAVITY AUTOENCODER FOR A NEURAL NETWORK WITH EMERGENT GRAVITY

To address the problem of quantum gravity from the AI-borne perspective of designing a holographic autoencoder, we need to first deal with the physics of machine learning and specifically enquire whether emergent quantum behavior can arise in a neural network. By emergent quantum mechanics, we mean a formulation within a framework of nonlocal hidden variables, as in the Bohm scheme [8]. Once emergent quantum behavior is shown to become possible within the machine learning scheme, we may address the question of developing a relativistic string gravitational scheme on the very same hidden variables adopted in the Bohm-like quantum network. Thus, the latter becomes in effect a quantum gravity autoencoder for the network representing emergent gravity.

To develop a network with emergent quantum behavior, we first need to focus on NNs in a statistical mechanics context, and closely examine the statistical physics underlying generic machine learning [9]. Thus, let us consider the NxN connectivity matrix w and the bias N-vector w_0 as stochastic variables with entries generically denoted q_i, $i = 1, ..., N + N^2$. The NN state vector x is thus updated in discrete time steps according to the usual scheme of f-activation presented in Chapter 1:

$$x(t+\tau) = f\left(wx(t) + w_0\right) \tag{6.1}$$

Here the time interval τ represents the overall thermalization time for the state vector x, whose entries will be regarded as the hidden variables. To develop the near-equilibrium thermodynamics scheme, let us define a loss function $J(x, q)$ that penalizes departures from the equilibrium which is achieved as $x(t + \tau) \approx x(t)$ for $t \gg \tau$. If μ represents the "reduced mass" of the network, we get

$$J(x,q) = \frac{1}{2}||x - f(wx + w_0)||^2 - \mu||x||^2 \tag{6.2}$$

Thus, the statistical thermodynamics near equilibrium stems directly from the canonical partition function

$$Z(\beta, q) = \int \exp\left[-\beta J(x, q)\right] d^N x \tag{6.3}$$

Following the tenets of statistical mechanics, the partition function yields the Helmholtz free energy for the NN given by:

$$A(\beta,q) = -\beta^{-1} \log\left[Z(\beta,q)\right] \tag{6.4}$$

In the specific case of an activation function given by the hyperbolic tangent, the partition function for the NN may be calculated as:

$$Z(\beta,q) = 2\pi^{N/2}\left\{\det\left[\beta G(w) + (1 - \beta\mu)I\right]\right\}^{-1/2}, \tag{6.5}$$

where $G(w)$ is given by [9]:

$$G(w) = (I - f'w)^T (I - f'w), \tag{6.6}$$

where f' is the diagonal matrix of first derivatives of the activation function with respect to each component of the NN state vector x.

When the network is at thermodynamic equilibrium, the average loss $\langle J(x, q)\rangle = U(\beta, q)$ becomes a minimum; hence, the Helmholtz free energy becomes

$$A(\beta,q) = \frac{1}{2}\beta^{-1}\left[-N\log(2\pi) + \sum_{\lambda_i > \beta^{-1}} \log(\lambda_i)\right] + \tilde{A}(\beta) \tag{6.7}$$

where λ_i are the eigenvalues of $G(w)$ and $\tilde{A}(\beta) = U(\beta) - \beta^{-1}S(\beta)$ is the thermodynamic Helmholtz free energy of the NN.

To describe the thermodynamic behavior of the NN near equilibrium, we take into account that the entropy production is stationary and introduce the time-dependent probability distribution $p(t, q)$ with Shannon entropy $S(t, q) = -\int p(t, q) \log [p(t, q)]dq$. Assuming the learning evolutionary drift follows (i.e., is proportional to) the gradient of the free energy, we get $\dfrac{dq_j}{dt} = \dfrac{\zeta\partial A}{\partial q_j}$; hence, the minimum entropy production yields the set of constraints

$$\frac{\partial A}{\partial t} + \zeta\left(\frac{\partial A}{\partial q_j}\right)^2 = \left\langle\frac{dA(t,q)}{dt}\right\rangle_t \tag{6.8}$$

This set of equations in turn begets a minimal action principle defined variationally as

$$\frac{\delta \mathcal{J}(p,A)}{\delta p} = \frac{\delta \mathcal{J}(p,A)}{\delta A} = 0, \tag{6.9}$$

where the action $\mathcal{J}(q,A)$ becomes

$$\mathcal{J}(p,A) = \int_0^\infty \frac{dS(t,q)}{dt} dt + \vartheta \int\int_{t=0}^\infty p(t,q)$$
$$\left\{ \frac{\partial A}{\partial t} + \sum_{j=1}^{N+N^2} \left[\zeta \left(\frac{\partial A}{\partial q_j} \right)^2 \right] - \left\langle \frac{dA(t,q)}{dt} \right\rangle_t \right\} dtdq \tag{6.10}$$

with ϑ denoting the corresponding Lagrange multiplier.

The action may be rewritten taking into account the Fokker–Planck equation satisfied by $p(t, q)$:

$$\frac{\partial p}{\partial t} = \frac{\partial}{\partial q_j} \left[\frac{D\partial p}{\partial q_j} - \frac{\zeta \partial A}{\partial q_j} p \right] \tag{6.11}$$

where the parameter D plays the role of diffusion coefficient for the NN dynamics near equilibrium. Taking into account Equation 6.11 in the computation of the time derivative of the Shannon entropy $S(t, q)$ enables us to rewrite Equation (6.10), so the action now reads [9]:

$$\mathcal{J}(p,A) = \int\int_{t=0}^\infty p(t,q)$$
$$\left\{ \vartheta \frac{\partial A}{\partial t} + \sum_{j=1}^{N+N^2} \left[\zeta \frac{\partial^2 A}{\partial q_j^2} - 4D \frac{\partial^2 p}{\partial q_j^2} + \vartheta \zeta \left(\frac{\partial A}{\partial q_j} \right)^2 \right] - \vartheta \left\langle \frac{dA(t,q)}{dt} \right\rangle_t \right\} dtdq \tag{6.12}$$

This action can be written equivalently in the form of a Schrödinger action:

$$\mathcal{J}(p,A) = \int\int_{t=0}^\infty \Psi^* \left[-4D \sum_{j=1}^{N+N^2} \frac{\partial^2}{\partial q_j^2} - i\eta \frac{\partial}{\partial t} + \Upsilon(q) \right] \Psi dtdq \tag{6.13}$$

with $\eta = \left(\dfrac{4D}{\zeta}\right)^{1/2}$, $\Upsilon(q) = -\left\langle \dfrac{dA(t,q)}{dt} \right\rangle_t$ and wave function

$$\Psi = p^{1/2} \exp\left[i\eta^{-1}A \right] \tag{6.14}$$

Thus, naming $\eta = \hbar$, we obtain the Schrödinger equation describing the state of the system as a particle-wave in the q-space $\mathbb{R}^{N \times N} \times \mathbb{R}^N$ of trainable variables for the learning process near equilibrium with state x-vector entries represented as thermalized hidden variables:

$$i\eta \dfrac{\partial}{\partial t} \Psi = \left[-4D\nabla^2 + \Upsilon(q) \right]\Psi \tag{6.15}$$

We have surveyed the statistical thermodynamics of an NN with stochastic trainable variables. The system evolves with the appropriate time coarse graining associated with the thermalization limit for hidden Bohm variables representing entries in the state vector. Such a system is capable of exhibiting an emergent quantum behavior. We are now ready to address the crucial question that underlies the quantum gravity conundrum from the standpoint of AI:

Can this quantum system be regarded as the variational autoencoder of an NN with emergent gravity in a Minkowski space-time?

To address this question, we need to consider the projection $\pi_\tau : x \rightarrow \bar{x}(q)$, where $\bar{x}(q)$ is the equilibrated state vector of the NN relative to the specific realization q of the stochastic trainable variables that in turn evolve in multiples of the equilibration time τ. In rigorous terms:

$$\bar{x}(q) = \int x \exp\left[-\beta J(x,q) \right] d^N x \tag{6.16}$$

Thus, we are enquiring whether it is possible to treat the nonequilibrium hidden variables (entries) in state vector x by mapping relativistic strings in an NN with an emergent Minkowski space, so that the entropy production in such a system is a function of the metric tensor that describes weak interactions between training subsystems of x-values. The answer is affirmative because Equations 6.5 and 6.6 may be specialized to the case where the weight vector w is simply a permutation matrix Ξ with an arbitrary number of cycles [9, 10], so that the matrix G now becomes

$$G = (\Xi - I)^T (\Xi - I). \tag{6.17}$$

Thus, the stochastic NN with partition function

$$Z(\beta,\Xi) = 2\pi^{N/2}\left\{\det\left[\beta(\Xi-I)^T(\Xi-I)+(1-\beta\mu)I\right]\right\}^{-1/2} \quad (6.18)$$

represents a quantum gravity autoencoder for the NN with emergent relativistic gravity, so that the following diagram becomes commutative:

$$x(t) \stackrel{\pi_\tau}{\rightarrow} \bar{x}(t,\Xi)$$

$$\downarrow F_{ST} \qquad \downarrow K_{QM}$$

$$x(t+\tau) \stackrel{\pi_\tau}{\rightarrow} \bar{x}(t+\tau,\Xi) \qquad (6.19)$$

where F_{ST} and K_{QM} represent, respectively, the string (Appendix, Section A.5) and quantum flow map.

6.3 RELATIVISTIC STRINGS-TURNED QUANTA IN MACHINE LEARNING PHYSICS

The statistical thermodynamics of machine learning is currently being elucidated by turning nontrainable (x) and trainable (q) variables into the stochastic variables for the NN and its variational autoencoder, respectively [9]. As demonstrated above, an NN may be endowed with emergent gravity while its autoencoder is governed by a latent Schrödinger Equation (6.15), thus exhibiting a quantum behavior. To generate this metamodel of quantum gravity, it is necessary to (a) treat the nontrainable variables as hidden variables in the emerging quantum gravity autoencoder, (b) consider a limit where the weight matrix (w) becomes a permutation matrix, and (c) treat the hidden variables in a nonequilibrium setting on timescales shorter than thermalization times by generating subsystems of state vectors whose dynamics are described by relativistic strings in an emergent Minkowski space-time (Appendix, Section A.5). The latent manifold associated with the Minkowski space-time is then obtained by thermalization of the hidden variables. The relativistic strings become enslaved or entrained in the thermalization limit where the nontrainable variables are treated as equilibrated vis-à-vis the trainable variables, and as such they are subsumed in the latent Schrödinger equation via the Helmholtz free energy.

Thus, we may conclude by stating that AI provides a quantum metamodel of gravity, and hence the big bang is in all likelihood a quantum event. In this context, at least one of the following four statements is valid:

A. The laws of quantum mechanics cannot be upheld in the big bang setting.

B. A quantum tunneling event generated the universe as progeny of another universe. The tunneling occurred across the barrier separating two quantum gravity autoencoders with respective latent manifolds Ω and Ω^*. The amplification of the information tunneled to the latent manifold Ω^* was realized as the funneled decoding of the latent wavefunction, giving rise to the progeny universe (see Figure 6.2).

C. The a priori presence of a primeval observer "Sensus Dei" materialized the big bang, which therefore is not a phenomenon-to-be in Wheeler's sense [11] but a realized event.

D. The big bang is a phenomenon-to-be in Wheeler's sense, and hence we are part of a multiverse, with the universe as possibility.

6.4 THE UNIVERSE AS A VARIATIONAL AUTOENCODER

This chapter addressed the conundrum of quantum gravity by reductively regarding the universe as the realization of a learning system with stochastic weights and biases where gravity and quantum behavior become emergent properties within the physics of machine learning. To that effect, we explore the possibility of an AI-based construction of a quantum holographic autoencoder which requires that the emergent quantum behavior arises in a neural network. By emergent quantum mechanics, we mean a formulation within a framework of nonlocal equilibrated hidden variables, as in the Bohm scheme [8]. Once an emergent quantum behavior is shown to become possible within the machine learning system equilibrated on the nontrainable – that is, hidden – variables, we address the question of developing a relativistic string gravitational scheme on the hidden variables adopted. Thus, the network with equilibrated nontrainable variables becomes in effect a quantum gravity autoencoder for the underlying network exhibiting emergent gravity in the nonequilibrium regime prior to the equilibration of the nontrainable variables (Appendix, Section A.5). In this way, we build a quantum metamodel for gravity that fulfills at least in part a major imperative for physicists seeking a unified field theory. Furthermore, the physical possibility of tunneling across quantum gravity autoencoders supports the idea that our universe may be the progeny of an older universe [12] that dreamt – or simulated – it.

This is a bold claim, yet it may be deconstructed vis-à-vis the main objective of this chapter, which was to describe the behavior of the neural networks in the limit where the bias vector, weight matrix, and state vector of neurons can be modeled as stochastic variables that undergo a learning evolution. As it turns out, this learning evolution, when projected onto the autoencoder, can be described by the time-dependent Schrödinger Equation (6.15), and the time evolution dictated by this equation is compatible with the relativistic decoding enshrined in the commutativity of the diagram presented in Equation (6.19). Taken together, these results have clear implications for the possible emergence of quantum mechanics, general relativity, and mesoscopic observers in neural networks governed by a unified theoretical scheme that adopts two different guises in the two different thermodynamic regimes (cf. Equation 6.19).

As it turns out, emergent quantum mechanics is a relatively new but rapidly evolving field based on a body of old and well-established ideas, dating back to the pilot wave theories of Louis de Brogie and David Bohm. The de Broglie–Bohm theory, also known as Bohmian mechanics, was originally formulated in terms of nonlocal hidden variables [8], which makes it quite straightforward to model an emergent quantum behavior in a stochastic neural network. Thus, our construct upholds the controversial view that quantum mechanics may not be a fundamental theory, but rather an *ansatz* giving rise to a mathematical tool which allows us to carry out statistical calculations with great efficacy and accuracy in a certain class of dynamical systems. In this guise, an emergent quantum mechanics should be derivable from the first principles of statistical mechanics. This is precisely what this chapter has accomplished for a dynamical system consisting of a neural network that is in effect a learning system that contains of two different types of degrees of freedom: the trainable bias vector and weight matrix elements and the nontrainable state vector of neurons, with the latter constituting the hidden variables.

Emergent (or we may say entropic) gravity is also a relatively new area of research, but in this case the picture is far more nebulous than in emergent quantum mechanics: It is far less clear whether progress has been made, if at all. The main hurdle is that emergent gravity requires also an emergent space, an emergent Lorentz invariance, and an emergent general relativity [13, 14]. To our surprise, the string-theory-based nonequilibrium treatment of the hidden variables in neural networks opened up a window of opportunity to treat the conundrum of emergent gravity in a completely unified fashion that encompasses all three aspects of the problem mentioned above in context of the learning dynamics. As it appears to be the case, a relativistic space-time can indeed emerge from a nonequilibrium evolution of the hidden variables in a manner that is very much akin to string theory [9]. More specifically, as described by Vanchurin [9], if one considers D minimally interacting subsystems (through bias vector and weight matrix) with average state vectors, then the emergent dynamics can be modeled with relativistic strings in an emergent D + 1 dimensional Minkowski space-time. Furthermore, the emergent dynamics may be modeled with the Einstein equations, provided the weak subsystem interactions are described by a metric tensor, as shown in the Appendix, Section A.5. In this way, a stochastic learning dynamics scheme such as the one proposed in this chapter proved to be instrumental for the equilibration of the emergent space-time that turned out to exhibit a behavior describable by a gravitational theory such as general relativity.

6.5 QUANTUM GRAVITY AUTOENCODERS AND THE ORIGIN OF THE UNIVERSE

The previous discussion addresses – but surely does not solve – one of the biggest problems concerning the history of the universe: What happened before the big bang? One cannot help but recall that Albert Einstein was never satisfied with the big bang scenario itself because he thought that a beginning in time would need to be postulated in a seemingly ad hoc manner. This way of doing physics to him was unacceptable, an indication of a feeble theory built on shaky grounds.

After nearly a century since Einstein voiced his skepticism – which extended to quantum mechanics itself – a whole gamut of hypothesis have been formulated regarding our cosmic origin. Perhaps the most sound include ideas such as the following: (a) the universe sprout from a quantum vacuum fluctuation, (b) the universe involves infinite cycles of contraction and expansion, (c) the universe was selected through the anthropic principle stemming from the string theory landscape of the multiverse, where every possible event or phenomenon is implicitly encompassed, none materialized, and the big bang itself is a phenomenon-to-be in Wheeler's sense, and (d) the universe emerged from the collapse of matter in the interior of a black hole that was contained in a progenitor universe.

Be as it may, none of these ideas can be ascribed full credibility, mainly because none of the theories they stem from has satisfactorily solved the conundrum of quantum gravity. A less explored and more daring possibility put forth in this chapter is that our universe was created in the laboratory by technologically advanced civilization capable of harnessing the power of quantum gravity autoencoders. This requires a mastery of the physics of learning machines and an ability to craft gravity and quantum behavior as emergent attributes of a stochastic learning system that admits a quantum gravity autoencoder (cf. Equation 6.19). The underlying theory behind this idea does not portend to solve the quantum gravity conundrum per se but at least reconciles the two main forces as emergent in a single learning machine. Furthermore, since our universe is endowed with a flat geometry at a zero net energy, an advanced civilization could have harnessed the power of quantum gravity autoencoders to create a baby universe through quantum tunneling [12] into a second quantum gravity autoencoder acting as reservoir for the spillover probability, as schematically depicted in Figure 6.2.

This "emergent matrix" idea of the origin of the universe reconciles the theological need for a "creator," that is, the primeval quantum observer that would have bestowed reality to the big bang by detecting the event,

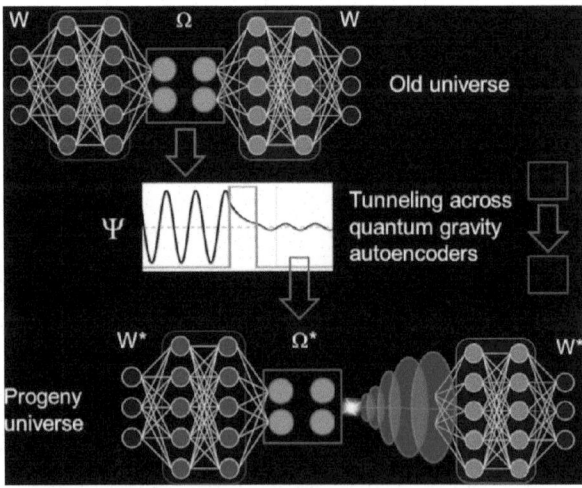

FIGURE 6.2   Schematic representation of a tunneling event across two quantum gravity autoencoders generating the universe (W^*, Ω^*) as progeny of an older universe (W, Ω). The amplification of information tunneled to the latent manifold Ω^* in a quantum spillover event materializes as the relativistic decoding of the latent wavefunction, a creation event giving rise to the progeny universe.

with the secular concept of quantum gravity. As said, while we have not gotten a cogent theory that conceptually unifies quantum mechanics and gravity, we have a "matrix framework of the universe," a neural network architecture where the two key forces in modern physics are reconciled, so that quantum behavior in the autoencoder can be decoded back as gravity through the holographic map that constitutes the inverse of the canonical projection. Surely, a more advanced civilization that masters the quantum gravity autoencoder technology would be able to accomplish the feat of creating baby universes leveraging quantum gravity autoencoders or other equivalent vehicles for quantum tunneling. If indeed it happened that our universe hosts such an advanced civilization that gave birth to a progeny universe by a mastery of the emergent physics in learning machines, then not only could it account for the matrix-like nature our universe, but it would also suggest that our universe behaves like a biological system that preserves its genetic blueprint through multiple generations. This advanced civilization would be able to leverage the quantum gravity autoencoders as birth channels for cosmic reproduction through quantum tunneling vehicles such as those described in Figure 6.2.

If so, our universe was not selected for us to dwell in it and bestow reality to the quantum events we are capable of detecting – as upheld by the standard anthropic principle – but rather, it was selected to host civilizations that are much more technologically advanced than we are. These civilizations capable of leveraging the technology required to create progeny universes, be it quantum gravity autoencoders or some unfathomable alternative vehicle efficacious at harnessing quantum tunneling, would be the actual drivers of the cosmic selection process.

By contrast, we are incapable at this time of harnessing technology for cosmological manipulation, and obviously we are incapable of recreating the cosmic conditions that led to our existence. In plain words, our civilization is cosmologically still at a rudimentary stage since we do not possess the technology to reproduce the universe that has hosted us for quite a while already. If we were to measure the technological level of a civilization by the ability to recreate or reproduce the astrophysical conditions that led to its existence, we would say that we are at early stages of development, possibly a low-level civilization, graded class C on a cosmic scale. By contrast, a civilization in the A class rank could recreate the astrophysical conditions that gave rise to its existence, namely produce a baby universe through a quantum-controlled laboratory experiment that leverages a tunneling effect through an appropriately crafted vehicle. A class A civilization would also be able to effectively address related challenges, such as producing a large enough density of dark energy to hold the universe together after its inception, as has already been discussed in the scientific literature [12].

While sobering and unpalatable, the assertion that human civilization is not a particularly smart one by measures such as the one proposed should hardly take us by surprise. Yet, if the universe proves to be a neural network with a learning physics capable of accommodating emergent quantum gravity, as argued in this chapter, we should not be surprised if we get upgraded to a rank A civilization in the near future.

6.6 TECHNOLOGIES FOR COSMOLOGICAL MANIPULATION LEVERAGING QUANTUM GRAVITY AUTOENCODERS

However sound and cogent, the theories on the quantum origins of the universe proposed so far [15] can be subject to a common and basic criticism: *There is no certainty that the universe can be treated as a quantum object.* This chapter heralds an improvement of this state of affairs as the duality enshrined in quantum gravity is realized through a purposely built autoencoder that compresses gravitational string data as emergent quanta.

In accord with Equation 6.15, if our universe is to become information-compressed within the quantum gravity autoencoder, the following relations must hold: $\zeta = \dfrac{1}{2m}$, $D = \dfrac{\hbar^2}{8m}$, or reciprocally:

$$\hbar = 2\sqrt{\frac{D}{\zeta}}, \qquad (6.20)$$

begetting the uncertainty relation adapted for the quantum gravity autoencoder:

$$\Delta E \times \Delta t \sim \sqrt{\frac{D}{\zeta}} \qquad (6.21)$$

While the total energy of the NN with emergent gravity is zero (a closed universe has zero energy [16]), the total energy of its quantum gravity autoencoder is $U = -\dfrac{\partial}{\partial \beta} \log Z(\beta, q) \neq 0$ in accord with the tenets of statistical mechanics. This paradox may be resolved by noting that the observational timescales for the gravitational NN and its autoencoder are different. Thus, in accord with the uncertainty principle, the total energy of the autoencoder may be zero for a timespan Δt which is incommensurably shorter than the equilibration time τ for the hidden variables. This implies that the parameter τ needs to be tuned as an architectural determinant of the autoencoder, so that the following relation is fulfilled:

$$\tau \gg \Delta t \sim \sqrt{\frac{D}{\zeta}} \left| \frac{\partial}{\partial \beta} logZ(\beta, q) \right|^{-1} \tag{6.22}$$

Thus, the incommensurability of the equilibration timescale relative to the timescales associated with the hidden variables modeled with relativistic strings implies that the universe may be a vacuum quantum fluctuation. This possibility is allowed by the uncertainty principle as described by Equation 6.22.

We should emphasize that the possibility that the universe as intelligible information is actually a vacuum quantum fluctuation is not as far-fetched as it may seem. A simple back-of-the-envelope calculation involving the cosmological constants of our known universe leads to an equivalent result. Thus, the energy $E = mc^2$ of a material object of mass m is actually counterbalanced by its gravitational potential energy $E_g = - GmM/R$, where G is the gravitational constant and M is the net mass of the universe contained within the Hubble ball of radius $R = c/H_0$, where H_0 is the Hubble constant [16]. To prove the previous assertion, we note that the critical minimal mass contained in the Hubble ball of volume $\frac{4}{3}\pi\left(\frac{c}{H_0}\right)^3$ and required for the universe to be closed is $M = c^3/(2GH_0)$, implying that the gravitational energy E_g compensates the energy E up to a constant of order unity.

The results described in this section pave the way for a cosmological technology that harnesses AI, or more specifically, the power of quantum gravity autoencoders. Thus, two parameters, D and τ, may be tuned to harness the power of AI to manipulate cosmological scales to the point of giving birth to a universe that serves as a metamodel for emergent quantum gravity.

REFERENCES

1. Maldacena J (1999) The Large N Limit of Superconformal Field Theories and Supergravity. *Int J Theor Phys* 38:1113.
2. Witten E (1998) Anti de Sitter Space and Holography. *Adv Theor Math Phys* 2:263–291.

3. Bekenstein JD (1973) Black Holes and Entropy. *Phys Rev D* 7:2333–2346.

4. Fernández A, Sinanoglu O (1982) The Lifting of an Inonu-Wigner Contraction at the Level of Universal Coverings. *J Math Phys* 23:2234.

5. Hashimoto K, Sugishita S, Tanaka A, Tomiya A (2018) Deep Learning and the AdS/CFT Correspondence. *Phys Rev D* 98:046019.

6. Matsueda H, Ishihara M, Hashizume Y (2013) Tensor Network and a Black Hole. *Phys Rev D* 87:066002.

7. Weinberg S (2008) *Cosmology*. Oxford University Press.

8. Bohm D (1962) A Suggested Interpretation of the Quantum Theory in Terms of Hidden Variables I. *Phys Rev* 86:166–179.

9. Vanchurin V (2021) Toward a Theory of Machine Learning. *Mach Learn Sci Technol* 2:036012.

10. Gubser SS (2010) *The little book of string theory*. Princeton University Press, Princeton, NJ.

11. Wheeler JA, Zurek WH (2014) *Quantum theory and measurement*. Princeton University Press, Princeton, NJ.

12. Farhi E, Guth A, Guven J (1990) Is it Possible to Create a Universe in the Laboratory by Quantum Tunneling? *Nuc Phys B* 339:417–490.

13. Vanchurin V (2018) Covariant Information Theory and Emergent Gravity. *Int J Mod Phys A* 33:1845019.

14. Bednik G, Pujolas O, Sibiryakov S (2013) Emergent Lorentz Invariance from Strong Dynamics: Holographic Examples. *J High Energy Phys* 11: 64.

15. Loeb A (2021) Was Our Universe Created in a Laboratory? Scientific American. October 15, 2021. https://www.scientificamerican.com/article/was-our-universe-created-in-a-laboratory/

16. Harrison ER (2003). *Masks of the universe*. Cambridge University Press, UK.

Epilogue

Φύσις κρύπτεσθαι φιλεῖ.

– HERACLITUS
(*Nature likes to keep secrets*; translation by the author)

E.1 TOPOLOGICAL METAMODELS BREED COMPUTATIONAL INTUITION

We now live in the era of Big Data, and we are building intelligent machines to process the data, shaping the field of data science. Today a whole range of technologies, from biomedical to astrophysical, are compelled to deal with daunting level of complexity enshrined in structures at a huge range of scales, from subcellular to cosmic. The data arising are often dynamic, that is, presented in time series, and the hope is that intelligent machines will team up with humans or perhaps work autonomously to discover the underlying models. To qualify as model, the scientific establishment is still demanding a significant dimensionality reduction and a set of coordinates that enable casting the dynamics in geometric terms, that is, as a sparse set of differential equations.

Given the complexities of the data that today's technologies must deal with, this program is likely to fail for at least two reasons: (a) we are applying the standards of seventeenth-century Newtonian physics to problems incommensurably more complex, noisy, and less structured than the ones dealt with in Newton's day and (b) we have not yet managed to build "intuition" in intelligent machines, so that most hierarchical structures in the data become unintelligible or undiscernible at a coarse level prior to a full burdensome feature extraction computation. This book addressed both problems squarely by harnessing state-of-the-art autoencoder technologies.

Because of its sheer complexity and the lack of structure, much of the data dealt with in the book is simply unyielding to the standard data analysis

DOI: 10.1201/9781003333012-9

tools available today. The big data dealt with in the book looks nothing like the kinds of data sets we would have encountered just a few years ago when the seventeenth-century paradigm was still upheld. The accepted paradigm hinged on the following operational tenets: Frame a hypothesis, decide which are the observables, make the measurement as accurately as possible, and distill the parsimonious model as a sparse set of differential equations in a latent lower dimensional space. As this book illustrates, today's big data stemming from a wide range of technologies is of very high dimensionality, dynamic, disorganized, uncurated, noisy, unstructured, and often incomplete, in other words not yielding to the accepted paradigm.

At the core of these issues, mathematician Ronald Coifman and his collaborators have become key players in the discussion [1]. Coifman maintains that a regression of today's big data requires the equivalent of a Newtonian revolution, a feat comparable with the invention of calculus. This book clearly upholds and contributes to cement this view. A revolution in data science is already underway, and in all likelihood topological metamodels may become the paragons of computational intuition.

In the spirit of the topological approach to computational intuition put forth in this book, mathematician Gunnar Carlsson has implemented a topological representation of complex big data sets [2]. Thus, the data endowed with a metric is laid onto graphs within manifolds that capture the similarity of the data points, so that point distances are mapped as relations within a topological object. In essence, this topology-based data analysis (TDA) attempts to circumvent the usual hurdles that plague the extraction of information from data sets that are very high-dimensional, incomplete, and noisy. TDA does so by adopting an approach based on "persistent homology" that is insensitive to the metric chosen to structure the data, while providing the desired dimensionality reduction and robustness to noise. Furthermore, the topological nature of Carlsson's approach endows it with a functorial character, a fundamental attribute that enables TDA to adapt to new mathematical tools.

As emphasized throughout, this book subscribes to the need for topological methodologies to unravel the structure of big dynamic data and guide model discovery. Nevertheless, the approach put forth in the book specifically differs from that of Carlsson or Coifman in that it exploits topological concepts that harness the power of autoencoder technology. Thus, our approach could be considered complementary to other efforts to implement intuitive topological modeling as we introduce a quotient space which becomes the latent space of the autoencoder.

E.2 AI PROBES AN EQUIVALENCE BETWEEN "WORMHOLE" AND QUANTUM ENTANGLEMENT

The examples from Chapters 3 to 6, from fields as distant as biomedicine and cosmology, highlight the power of topology-based metamodels in leveraging autoencoder technology for model discovery. The main assertion in Chapter 6 is that a neural network with emergent gravity admits an autoencoder with emergent quantum behavior and hidden variables that can be modeled with relativistic strings. Thus, a quantum metamodel of gravity is in principle possible. This finding naturally paves the way to tackle the next conundrum involving the conjectured equivalence between relativistic wormholes and quantum entanglement.

Space-time locality is a basic tenet of modern physics. The term "locality" refers to the impossibility of sending signals at speeds higher than the speed of light, an idea that is constantly challenged both by quantum mechanics and by general relativity. Thus, quantum mechanics gave rise to the Einstein–Podolsky–Rosen (EPR) correlations also termed "entanglements," while general relativity allows for solutions to the equations of motion that connect distant regions through short-circuiting "wormholes" also known as Einstein–Rosen (ER) bridges [3]. Physicists Maldacena and Susskind have conjectured that these two concepts may be connected by a duality that becomes in effect equivalence, akin to the similar duality found in the physical underpinnings of quantum gravity [3]. They have argued persuasively that the ER bridge between two black holes may be actually created by EPR correlations between microstates of the two black holes, and labeled the conjecture "ER = EPR relation." In their analysis, the ER bridge is a type of EPR correlation in which the correlated quantum systems are in a specific entangled state that admits a weakly coupled Einstein gravity description. This situation is illustrated by a black hole pair creation in a magnetic field, and it is tempting to think that any EPR correlated system, even a simple singlet state of two spins, is connected by some sort of ER bridge.

The neural network model of the universe endowed with the quantum gravity autoencoder described in Chapter 6 seems an ideal system to validate (or disprove) the EPR=ER relation. This is because the emergent quantum behavior of the autoencoder pivots on hidden variables that are interacting through relativistic strings and the very existence of the hidden variables is known to resolve the EPR paradox [4]. In the quantum gravity autoencoder, the trainable variables conforming the $\mathbf{q}$-vector exhibit a quantum mechanical behavior in an equilibrium regime where the network state variables that constitute the **x**-vector have been thermalized. A learning process involves L separate sets of training **x**-vectors with expected values $\bar{\boldsymbol{x}}^l, l = 1, 2, \ldots, L$ and the expectation vectors together with the expectation state vector $\bar{\boldsymbol{x}}^0$, representing network evolution to zeroth order in the linear approximation to node activation, are regarded as the hidden variables in the emergent quantum behavior of the trainable **q**-states. On the other hand, the nonequilibrium dynamics of the hidden variables becomes relevant on timescales much smaller than their thermalization time. This nonequilibrium dynamics are determined by the strength of the weak interactions between vector pairs $\bar{\boldsymbol{x}}^\nu, \bar{\boldsymbol{x}}^\xi, \nu, \xi = 0, 1, \ldots, L$ quantified by the tensor $g^i_{\nu\xi}$, where the dummy index i labels each neuron in the system. By weak interactions, we mean that the generic vectors $\bar{\boldsymbol{x}}^\nu$ and $\bar{\boldsymbol{x}}^\xi$ are not interacting directly but through the **q**-vector that they themselves contribute to train. These dynamics can be cast in terms of relativistic strings in an emergent space-time, as indicated in Chapter 6.

This AI framework with emergent physics seems ideal to validate the EPR=ER relation. If two parts of the **q**-space are entangled in the quantum autoencoder, they should be bridged through a wormhole in the full network with emergent gravity, as described in Figure E.1. The presumed proportionality between the gravitational action and the quantum action may be a good starting point for this project aimed at testing the power of

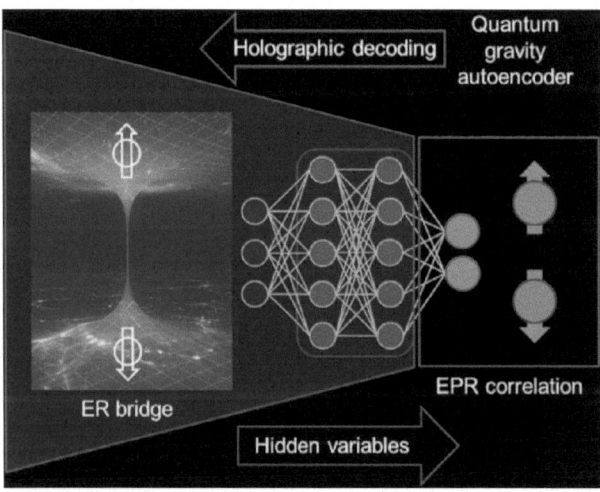

FIGURE E.1   Schematics of quantum gravity autoencoder (Chapter 6) deployed to validate the "EPR=ER relation."

AI as purveyor of the physical underpinnings for the most perplexing problems in modern cosmology.

E.3 EMERGENT SPACE-TIME TOPOLOGY CREATED BY ENTANGLING QUANTUM-GRAVITY AUTOENCODERS

The narrative in this book takes an unexpected turn in Chapter 6, where it is shown that a connected array of neurons (NN) may be treated not only as an information processing machine but also as a statistical mechanical object, capable of exhibiting emergent physical behavior. Thus, a duality between emergent gravity and quantum mechanics is established through an autoencoder that thermalizes hidden degrees of freedom arrayed on the NN state vector $\mathbf{x}$. In this way, the autoencoder exhibits emergent quantum behavior while the nonequilibrium dynamics of the hidden variables spans an emergent space-time endowed with gravity (Chapter 6 and Appendix, Section A.5). This physical duality of the learning machine cast in terms of statistical mechanics should enable topological innovation on the emergent space-time through quantum entanglement at the autoencoder level.

To illustrate this point, consider two completely separated replicas of a quantum gravity autoencoder (QGAE), labeled left (L) and right (R) with identical collection of training sets $\left\{ \mathcal{S}^{(\mu)} \right\}_\mu$ comprised of $\mathbf{x}$-vectors and generating the sets of $\mathbf{q}$-vectors $Q_L = \left\{ Argmin\, J_<^{(\mu,L)} \right\}_\mu$, $Q_R = \left\{ Argmin\, J_<^{(\mu,R)} \right\}_\mu$, in QGAE(L) and QGAE(R), respectively (Appendix, Section A.5). Notice that the stochastic nature of the training process implies that $Argmin\, J_<^{(\mu,L)}$ is not necessarily equal to $Argmin\, J_<^{(\mu,R)}$, but surely the $\mathbf{q}$-vector pairs $Argmin\, J_<^{(\mu,L)}$, $Argmin\, J_<^{(\mu,R)}$ are entangled as they have a common origin. Hence, if the NNs are architecturally configured so that there is a black hole in Q_L and therefore in Q_R, then both black holes are necessarily entangled. This entanglement begets connectivity in the respective emergent space-times associated with QGAE(L) and QGAE(R), which would be topologically related to the Penrose diagram shown in Figure E.2 (cf. [3]).

Thus, the equivalent of a wormhole or Einstein–Rosen (ER) bridge between the black holes would be sustained between the two learning machines L and R as a result of entanglement at the QGAE level. This "double black hole" would have two exteriors (L and R), and two event horizons meeting at the counterpart of the ER bridge shown in the Penrose diagram (Figure E.2). However, we expect this sort of connectivity to be

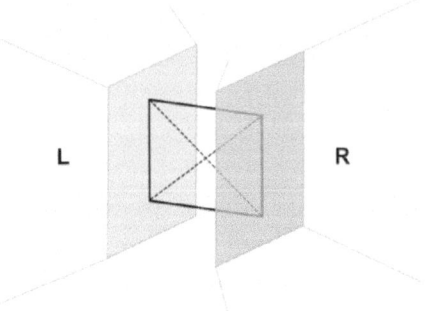

FIGURE E.2 Penrose diagram counterpart for the connection topology in emergent space-time arising from the entanglement of two black holes within quantum gravity autoencoders (L and R). The respective event horizons are marked by dashed-line upper and lower triangles in a conformal two-dimensional space-time.

topologically different in the case of entangled NNs. This is so because the emergent space-time constructed upon the q-space of a QGAE is essentially different from the Minkowski space where the particular solution to Einstein's equations was obtained and topologically described by the Penrose diagram.

Describing the space-time topology of entangled NNs may prove rewarding, as it is likely to herald a new breed of quantum computation that we may provisionally term "quantum gravity computation."

The proposals put forth in Sections E.2 and E.3 should mitigate at least in part the anxiety brewed by the question that always arises as a book comes to an end: Where do we go from here?

E.4 METAMODEL DISCOVERY WITH EXTRA DIMENSIONS AND LOST SYMMETRIES: ARTIFICIAL INTELLIGENCE DECONSTRUCTS QUANTUM MECHANICS

This book explored the concept of metamodels of physical phenomena as a means to develop intuition in AI systems that are leveraged to discover models. The metamodel construction pivots on the possibilities of generalized autoencoders to capture and propagate a simplified but decodable version of a dynamical system in a latent space. This operation typically entails a significant dimensionality reduction. As previously shown (Chapters 1 and 2), the latent space or latent manifold may be mathematically identified with a quotient space. Thus, a conceptual relation may be drawn between autoencoder and quotient space.

Invariably, when we think of discovering the physical model that underlies a dynamical system, we think of dimensionality reduction, of retaining the "essential" variables, of averaging out degrees of freedom and removing noisy sources. Metamodeling further enhances this tendency to the point that the latent dynamics can no longer be cast in terms of differential equations but are ruled by combinatorics. The metamodel is at the same time more elaborate and also more primitive than the model, akin to a Picasso painting from his later days compared with his early work. But not all modeling efforts necessarily entail dimensionality reduction. The daunting difficulties with quantum mechanics calculations suggest that maybe things would look simpler if we incorporate extra dimensions and recover lost symmetries. Of course, that is easier said than done. What would such dimensions and symmetries be? Our understanding is not yet ripe to draw definitive conclusions.

More mundane examples suggest that modeling by incorporating extra dimensions and lost symmetries may represent a viable and reasonable approach. For instance, physicist Lisa Randall focused on quasi-crystals, whose inherent order only becomes apparent when extra dimensions are incorporated [5]. At first glance, quasi-crystals may look organized, but upon close examination we notice they lack the precise regularity that is the hallmark of a crystal. In fact, quasi-crystals may be viewed as projections of higher dimensional crystalline structures. Full symmetry is apparent not in our well-worn spatial dimensions but in a higher dimensional space [5]. A Penrose tiling in two dimensions is a good avatar for this type of regularity, where the lost symmetry can only be recovered when the pattern is regarded as a projection of a five-dimensional crystalline pattern (Figure E.3).

When we think of physical space, or rather a space where we can do physics, we cannot a priori decide whether the space is actually a quotient space, and hence we have "lost" symmetries, or it is the ultimate universal covering. If we admit for a moment, as Chapter 6 suggests, that the universe is a neural network with emergent physics, we would be confronted with the problem of whether the universe is actually an autoencoder or a neural network that admits an autoencoder when some superior entity endeavors to model it. This is because the question of whether physical space may be a quotient space associated with lost symmetries cannot be decided a priori: it would require the design of new tests that may potentially be discovered with the assistance of AI. In fact, there are two key properties of the "lost symmetry group" (G) that would enable the

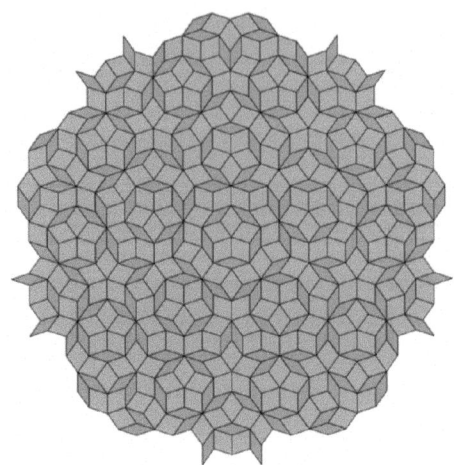

FIGURE E.3   Penrose tiling as encoder of lost symmetry and avatar of a quasi-crystal: An aperiodic pattern resulting from the projection in two dimensions of a five-dimensional crystalline structure. *Inductiveload*, the copyright holder of this work, released it into the public domain (https://commons.wikimedia.org/w/index.php?curid=5839079).

quotient space W/G to inherit the metric of its universal covering W: (a) G has no fixed points ($\forall g \in G, g \neq I, \forall x \in W : gx \neq x$) and (b) G is "properly discontinuous," an attribute that implies the fulfillment of the following condition:

$$\forall x \in W, \exists\, open\; ball\; U_x \ni x : \forall g_1, g_2 \in G, g_1 \neq g_2, g_1 U_x \cap g_2 U_x = \varnothing.$$

It is often emphasized that the subatomic world is divorced from macroscopic experience because it has discretized observables, as reflected by the discrete quantum numbers. However, in light of this discussion we dare say this discretization, a hallmark of quantum mechanics, may be in a sense artifactual and removable altogether if we chose to operate at the level of a universal covering and incorporate extra dimensions [6]. To shed some light on this assertion, let us consider the following illustration from early-days atomic physics: Assume a unitary radius for the Bohr electron orbit, then the de Broglie wavelength λ is forced to adopt discrete values according to $n\lambda = 2\pi$, $n \in \mathbb{N}$, in order to prevent destructive out-of-phase interference on the stationary matter wave that represents the electron. However, the unitary circle S^1 is isomorphic to the quotient space $\mathbb{R}/Z$ for a single space dimension with a lost translational symmetry associated

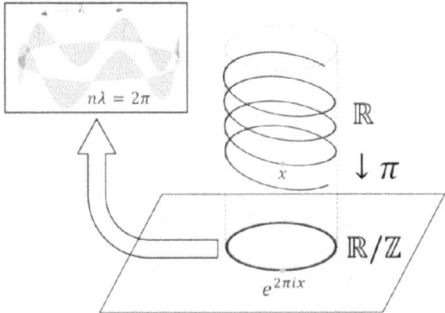

FIGURE E.4 Incorporating lost symmetry to "simplify" quantum mechanics at the level of universal coverings. This decoding is likely to challenge the assumed discrete nature of subatomic variables.

with the group of integers (Figure E.4). But within the universal covering $\mathbb{R}$ of $S^1 \approx \mathbb{R}/Z$ there is no quantization, as the wavelength is not constrained to fulfill any condition to avoid destructive phase interference.

This argument suggests that the discrete nature of observables in the quantum world may be simply the result of a lost symmetry arising in a universal covering of which we are not aware. Like in the quasi-crystal, the structure is only fully revealed when extra dimensions are incorporated. The problem would be akin to decode the latent manifold associated to an autoencoder that everyone believed to be the dynamical system itself in order to generate the actual dynamical system. Such an endeavor stands at the antipodes of the approaches introduced in this book for model discovery, where the dichotomy reality model is never challenged. We are proposing to consider quantum mechanics itself as a model resulting from a projection or dimensionality reduction and seek to discover the physical system that projects onto the latent manifold in the model. As it happens, the latent manifold required quantization, which suggests that a discretized parametrization of the dynamics must hold for the subatomic world. But we now turned the tables and claim that this is simply the way the autoencoder discovered the model. Thus, our interests have shifted as we endeavor to decode the latent dynamics associated with the autoencoder into a system that operates at the level of its universal covering.

In efforts to simplify quantum mechanics as applied to transformations of matter resulting in chemical reactions, a metamodel has been developed. This metamodel includes concepts like atomic and molecular orbitals and combinatorial rules, such as the conservation of orbital symmetry. Such metamodels have added considerable predictive value to the field of quantum chemistry, especially in areas like physical organic chemistry [7], and allowed researchers to circumvent the wanton difficulties in trying to solve the time-dependent Schrödinger equation for such complicated systems. In accord with our proposal of incorporating extra dimensions and lifting quantum mechanics to the level of a universal covering, we cannot help wondering what would the current quantum metamodel turn into (Figure E.5). What would be the counterparts of molecular and atomic orbitals and orbital symmetry combinatorics? The inherent nature of quantum mechanics as it now stands enabled such combinatorial rules and shorthand simplifications. It is likely that the incorporation of extra dimensions may render the counterpart of this metamodel simpler or perhaps even superfluous.

This proposal may well make quantum mechanics more accessible computationally but its ultimate relevance is contingent on identifying

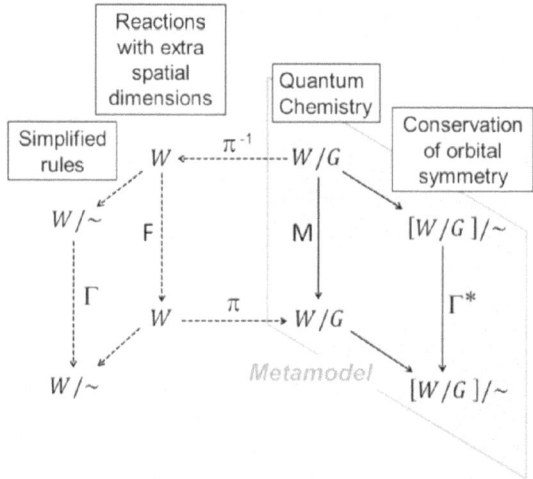

FIGURE E.5 Lifting of the quantum mechanics at the level of universal covering by incorporating lost symmetries and extra spatial dimensions. Vertical arrows denote events at specific levels of description, while horizontal arrows denote projections or their inverses. Quantum mechanics as it stands is here assumed to operate on the quotient space W/G, with a "lost" symmetry group G of whose existence we are currently oblivious. The computational complexity of quantum mechanics as it now stands demanded the creation of a metamodel that significantly enhanced its predictive value in areas like quantum organic chemistry. A lifting of quantum mechanics at the level of universal coverings in the spirit of Figure E.4 is likely to simplify calculations considerably and may render metamodeling simpler or even superfluous.

clues on the lost symmetries and overlooked dimensions of space. This is a goal where the assistance of AI is likely to be instrumental since an auto-encoder will drive the modeling effort in the first place.

A problem with the extra dimension proposal pertains to the way in which it may reveal itself in the standard four-dimensional space-time. The energy of a particle in the 3D space is made up of the rest mass energy, given by $E = mc^2$, and the kinetic energy associated with its motion in 3D. If the particle has the freedom to move in the extra dimension, its extra kinetic energy would be perceived and interpreted as part of the rest energy, or equivalently, as contributing to the mass of the particle. Thus, particles with a translational degree of freedom along the extra dimension should have in fact a larger apparent mass than those catalogued in the Standard Model, which are assumed to be at rest along the extra dimension. Thus, yet undiscovered heavier versions may exist for all particles in the Standard Model. If motion is quantized along the extra dimension, then we should expect hierarchies of versions of each particle, one version for each value of the denumerable extra quantum number. Such hierarchies would be akin to the so-called Kaluza-Klein (KK) towers [5].

This apparent mass increase associated with kinetic energy along the hidden dimension is already playing out in current speculation on the origin of "dark matter," a culprit for the massive gravitational pull that appears to hold together the structure of the universe and made up of particles not detectably interactive with photons. According to the scheme previously presented (Figure E.4), an electron, and a priori any particle in the Standard Model, may be "dark" if the motion that involves the extra dimension projects onto a wave with destructive phase interference in the 3D space. According to the discoverable AI model, the electron is always moving in the universal covering of its 3D projection rather than on its perceived/detectable latent space. This means that only motion involving the extra dimension capable of generating a quantized stationary 3D projection would yield a "visible" particle. We may envision "dark" particles as those rendering destructive phase interference in the standard 4D space-time. In other words, we envision particles that only exist as field excitations when the extra dimension is incorporated in the decoded universal covering but would not be part of ordinary matter as they would be undetectable in the 4D space-time. On the other hand, it would not be possible to radiate a detectable electron and "steer it into darkness" simply because photons do not interact along the extra dimension.

This closing argument prompts the following questions: Would it be possible that most particles in the Standard Model have "dark" versions along the lines described? How can we find evidence supporting the existence of such dark particles? The methods described in this book may shed light on such fundamental problems, as we count on the awareness that 4D space-time is a latent space identified by an autoencoder, as opposed to the actual space containing the training and testing data. Thus, field excitations in four spatial dimensions would likely yield additional particles not present in the ordinary three spatial dimensions, as novel oscillatory modes arise and become endowed with mass via Yukawa coupling with an ur-Higgs field (fermions) or via potential energy terms arising from covariant derivatives in the kinetic energy of the ur-Higgs field (bosons). Within this informational framework, decoding the Lagrangian dynamics of the ur-Higgs-field-particle combinations in the latent manifold becomes perhaps the way to unravel one of the biggest mysteries surrounding the sustainability of the structure of the universe. The lifting of these Lagrangians to the level of universal covering incorporating extra dimensions is likely to yield unfathomable dark matter particles. Ultimately, the proof that AI is capable of decoding/deconstructing the Standard Model would require that the particle masses bestowed by the Higgs field can be predicted based on the particle wave functions projected on the extra dimensions. May this be the subject for the next book!

REFERENCES

1. Sroczynski DW, Kemeth FP, Coifman RR, Kevrekidis IG (2022) Questionnaires to PDEs: From Disorganized Data to Emergent Generative Dynamic Models. arXiv:2204.11961 [math.DS].
2. Carlsson G (2009) Topology and Data. *Bull Amer Math Soc* 46:255–308.
3. Maldacena J, Susskind L (2013) Cool Horizons for Entangled Black Holes. *Fortschr Phys* 61:781–811.
4. Bohm D, Aharonov Y (1957) Discussion of Experimental Proof for the Paradox of Einstein, Rosen and Podolsky. *Phys Rev* 108:1070–1079.
5. Randall L (2006) *Warped passages: Unraveling the mysteries of the universe's hidden dimensions.* Ecco Press, New York.
6. Fernández A, Sinanoglu O (1982) The Lifting of an Inonu-Wigner Contraction at the Level of Universal Coverings. *J Math Phys* 23:2234–2235.
7. Fernández A (1985) 1,3-Sigmatropic Thermal Rearrangements as Vector Fields on the 2-Sphere. *J Chem Phys* 82:3123–3128.

Appendix

A.1 CODE FOR DEHYDRON IDENTIFICATION

The software "YapView" (yet another protein view), currently incarnated as "Dehydron Calculator," is used to identify the dehydrons from structural coordinates of soluble proteins. This program and two equivalent programs, "WRAPPA" and "dehydron calculator in PyMOL plugin version," are currently freely downloadable from the site:

http://people.cs.uchicago.edu/~ridg/softwaredigbio.html

The code for the Dehydron Calculator in Python provided as plugin for PyMol is given in this Appendix. The open source is found at:

https://raw.github.com/Pymol-Scripts/Pymol-script-repo/master/plugins/dehydron.py

YAP View/Dehydron Calculator and the "desolv" plugin within it used to calculate hydrogen bond burial and identify dehydrons are currently hosted at:

http://sourceforge.net/projects/protlib/files/yapview/0.6.8/

The main installer for the windows version (YAPView-0.6.8-1-Installer. exe) comes with a precompiled version of the desolv plugin that should be accessible from the UI (user interface).

Code for Dehydron Calculator as PyMOL Plugin (in Python)

```
Dehydron: A dehydron calculator plugin for PyMOL
Version: 1.7
Described at PyMOL wiki:
http://www.pymolwiki.org/index.php/dehydron

Author : Osvaldo Martin
email: aloctavodia@gmail.com
Date    : March 2013
License: GNU General Public License
```

Acknowledgement: The H-bond detection code is based on
  the list _ mc _ hbonds.py
script from Robert L. Campbell
http://pldserver1.biochem.queensu.ca/~rlc/work/pymol/
'''

```
import Tkinter
from Tkinter import *
import Pmw
from pymol import cmd
from pymol import stored

def ___ init ___ (self):
    """Add this Plugin to the PyMOL menu"""
    self.menuBar.addmenuitem('Plugin', 'command',
                             'dehydron',
                             label = 'Dehydron',
                             command = lambda :
mainDialog())

def mainDialog():
    """ Creates the GUI"""

    def get _ dehydrons():
        angle _ range = float(angle _ value.get())
        max _ distance = float(dist _ cutoff _ value.get())
        desolv = float(desolv _ sphere.get())
        min _ wrappers = float(min _ value.get())
        selection = sel _ value.get()
        dehydron(selection, angle _ range, max _ distance,
desolv, min _ wrappers)

    master = Tkinter.Tk()
    master.title(' dehydron ')
    w = Tkinter.Label(master, text = 'dehydron
calculator\nOsvaldo Martin - aloctavodia@gmail.com',
                                background = '#000000',
                                foreground = '#cecece',
                                #pady = 20,
                                )
    w.pack(expand=1, fill = 'both', padx = 4, pady = 4)

    Pmw.initialise(master)
    nb = Pmw.NoteBook(master, hull _ width = 420,
hull _ height=280)
    p1 = nb.add('Main')
    p2 = nb.add('About')
```

```
    nb.pack(padx=5, pady=5, fill=BOTH, expand=1)
########################### Main TAB ##################
##############
### hydrogen bond settings
    group = Pmw.Group(p1,tag_text='Hydrogen bond Settings')
    group.pack(fill='x', expand=1, padx=20, pady=1)
    Label(group.interior(), text='angle range').grid(row=2,
column=0)
    angle_value = StringVar(master=group.interior())
    angle_value.set(40)
    entry_angle = Entry(group.interior(),textvariable=
angle_value, width=10)
    entry_angle.grid(row=2, column=1)
    entry_angle.configure(state='normal')
    entry_angle.update()
    Label(group.interior(), text='max distance').
grid(row=3, column=0)
    dist_cutoff_value = StringVar(master=group.
interior())
    dist_cutoff_value.set(3.5)
    entry_dist =
Entry(group.interior(),textvariable=dist_cutoff_value,
 width=10)
    entry_dist.grid(row=3, column=1)
    entry_dist.configure(state='normal')
    entry_dist.update()
### dehydron settings
    group = Pmw.Group(p1,tag_text='Dehydron Settings')
    group.pack(fill='x', expand=1, padx=20, pady=5)
    Label(group.interior(), text='desolvatation sphere
radius').grid(row=2, column=2)
    desolv_sphere = StringVar(master=group.interior())
    desolv_sphere.set(6.5)

entry_desolv=Entry(group.interior(),textvariable=
desolv_sphere, width=10)
    entry_desolv.grid(row=2, column=3)
    entry_desolv.configure(state='normal')
    entry_desolv.update()
    Label(group.interior(), text='minimum wrappers').
grid(row=3, column=2)
    min_value = StringVar(master=group.interior())
    min_value.set(19)
    entry_min_value=Entry(group.interior(), text
variable=min_value, width=10)
    entry_min_value.grid(row=3, column=3)
```

```
    entry _ min _ value.configure(state='normal')
    entry _ min _ value.update()
### selection settings
    group = Pmw.Group(p1,tag _ text='Selection')
    group.pack(fill='x', expand=1, padx=20, pady=5)
    Label(group.interior(), text='selection').grid
(row=4, column=2)
    sel _ value = StringVar(master=group.interior())
    sel _ value.set('all')
    entry_sel_value=Entry(group.interior(),text
variable=sel _ value, width=10)
    entry _ sel _ value.grid(row=4, column=3)
    entry _ sel _ value.configure(state='normal')
    entry _ sel _ value.update()
### submit
    Button(p1, text="Calculate", command=get _ dehy-
drons).pack(side=BOTTOM)
############################ About TAB ################
###############
    group = Pmw.Group(p2, tag _ text='About dehydron
plug-in')
    group.pack(fill = 'both', expand=1, padx = 5, pady = 5)
    text =u"""For a brief introduction to the dehydron
concept, you could
read http://en.wikipedia.org/wiki/dehydron

Citation for this plugin:
Martin O.A.; dehydron calculator plugin for PyMOL,
2012. IMASL-CONICET.

Citation for PyMOL may be found here:
http://pymol.sourceforge.net/faq.html#CITE

Citation for dehydrons (I think these could be used):
Fern\u00E1ndez A. and Scott R.; "Adherence of packing
defects in soluble proteins", Phys. Rev. Lett. 91, 018102
(2003).

Fern\u00E1ndez A., Rogale K., Scott R. and Scheraga H.A.;
"Inhibitor design by wrapping packing defects in HIV-1
proteins", PNAS, 101, 11640-45 (2004).

Fern\u00E1ndez A. "Transformative Concepts for Drug
  Design:
Target Wrapping" (ISBN 978-3642117916),
Springer-Verlag, Berlin, Heidelberg (2010).
"""
```

```
    #
    # Add this as text in a scrollable panel.
    # Code based on Caver plugin
    # http://loschmidt.chemi.muni.cz/caver/index.php
    #
    interior _ frame = Frame(group.interior())
    bar = Scrollbar(interior _ frame)
    text _ holder = Text(interior _ frame,
yscrollcommand=bar.set, foreground="#cecece",background=
"#000000",font="Times 12")
    bar.config(command=text _ holder.yview)
    text _ holder.insert(END,text)
    text _ holder.pack(side=LEFT,expand="yes",fill=
"both")
    bar.pack(side=LEFT,expand="yes",fill="y")
    interior _ frame.pack(expand="yes",fill="both")

    master.mainloop()

def dehydron(selection='all', angle _ range=40, max _
  distance=3.5, desolv=6.5, min _ wrappers=19, quiet=0):
    '''
DESCRIPTION

    dehydron calculator

USAGE

    dehydron [ selection [, angle _ range [, max _
distance [, desolv [, min _ wrappers ]]]]]
    '''
    angle, max _ distance = float(angle _ range),
float(max _ distance)
    desolv, min_wrappers = float(desolv) , int(min_
wrappers)
    quiet = int(quiet)

    name = cmd.get _ legal _ name('DH _ %s' % selection)
    cmd.delete(name)

    selection _ hb = '((%s) and polymer)' % (selection)
    hb = cmd.find _ pairs("((byres "+selection _ hb+") and
n. n)","((byres "+selection _ hb+") and n.
o)",mode=1,cutoff=max _ distance,angle=angle _ range)
```

```
    if not quiet:
        hb.sort(lambda x,y:(cmp(x[0, 1],y[0, 1])))
        print "----------------------------------------
-----------------------"
        print "-------------------------Dehydron Results---
---------------------"
        print "----------------------------------------
-----------------------"
        print "                    Donor               |
Aceptor              |"
        print "      Object    Chain Residue   |
Object    Chain Residue   | # wrappers"

    cmd.select('_nonpolar', '(elem C) and not (solvent
or (elem N+O) extend 1)', 0)
    try:
        cmd.select('_selection', '%s' % selection, 0)
    except:
        pass

    sel = []
    for pairs in hb:
        wrappers = cmd.count_atoms('((%s and _nonpolar
and _selection) within %f of byca (%s`%d %s`%d))' %
                        ((pairs[0][0], desolv) + pairs[0] +
pairs[1]))
            if wrappers < min_wrappers:
                cmd.distance(name, pairs[0], pairs[1])
                if not quiet:
                    cmd.iterate(pairs[0], 'stored.nitro =
chain, resi, resn')
                    cmd.iterate(pairs[1], 'stored.oxy =
chain, resi, resn')
                    print ' %12s%4s%6s%5d | %12s%4s%6s%5d
|%7s' % (pairs[0][0], stored.nitro[0], stored.nitro[2],
int(stored.nitro[1]), pairs[0, 1], stored.oxy[0], stored.
oxy[2], int(stored.oxy[1]), wrappers)
                sel.append(pairs[0])
                sel.append(pairs[1])
    cmd.delete('_nonpolar')
    cmd.delete('_selection')

    if len(sel) > 0:
        cmd.show_as('dashes', name)
    elif not quiet and len(hb) != 0:
        print ' - no dehydrons were found - '
    else:
```

```
        print ' - no hydrogen bonds were found - '
cmd.extend('dehydron', dehydron)

# vi:expandtab:smarttab
```

NOTE: The code also runs directly from the line of command. It is basically the same as the plugin but adapted to be called from a terminal. When executed as python dehydron_ter.py > log.out, it will compute dehydrons for all PDB files contained in the same folder as "dehydron_ter. py," and it would download the results in the log.out file. The code can of course be modified to print the results on a single file (trivial in Python).

```
import ___ main ___
___ main ___ .pymol _ argv = ['pymol','-qck']
import glob
import pymol
from pymol import cmd, stored
pymol.finish _ launching()

'''
Dehydron: A dehydron calculator plugin for PyMOL
Version: 1.7
Described at PyMOL wiki:

http://www.pymolwiki.org/index.php/dehydron
'''
def dehydron(selection='all', angle _ range=40, max _
distance=3.5, desolv=6.5, min _ wrappers=19, quiet=0):
    '''
DESCRIPTION

    dehydron calculator

USAGE

    dehydron [ selection [, angle _ range [, max _ dis-
tance [, desolv [, min _ wrappers ]]]]]
    '''
    angle, max _ distance = float(angle _ range),
float(max _ distance)
    desolv, min _ wrappers = float(desolv), int(min _
wrappers)
    quiet = int(quiet)

    name = cmd.get _ legal _ name('DH _ %s' % selection)
    cmd.delete(name)
```

```
    selection _ hb = '((%s) and polymer)' % (selection)
    hb = cmd.find _ pairs("((byres "+selection _ hb+") and
n. n)","((byres "+selection _ hb+") and n.
o)",mode=1,cutoff=max _ distance,angle=angle _ range)

  if not quiet:
      hb.sort(lambda x,y:(cmp(x[0, 1],y[0, 1])))
      print "---------------------------------------------
----------------------"
      print "------------------------Dehydron Results
------------------------"
      print "---------------------------------------------
----------------------"
      print "              Donor             |
Aceptor           |"
      print "      Object   Chain Residue    |
Object   Chain Residue   | # wrappers"

  cmd.select(' _ nonpolar', '(elem C) and not (solvent
or (elem N+O) extend 1)', 0)
  try:
      cmd.select(' _ selection', '%s' % selection, 0)
  except:
      pass

  sel = []
  for pairs in hb:
      wrappers = cmd.count _ atoms('((%s and _ nonpolar
and _ selection) within %f of byca (%s`%d %s`%d))' %
              ((pairs[0][0], desolv) + pairs[0] +
pairs[1]))
      if wrappers < min _ wrappers:
          cmd.distance(name, pairs[0], pairs[1])
          if not quiet:
              cmd.iterate(pairs[0], 'stored.nitro =
chain, resi, resn')
              cmd.iterate(pairs[1], 'stored.oxy =
chain, resi, resn')
              print ' %12s%4s%6s%5d | %12s%4s%6s%5d
|%7s' % (pairs[0][0], stored.nitro[0], stored.nitro[2],
int(stored.nitro[1]), pairs[0, 1], stored.oxy[0], stored.
oxy[2], int(stored.oxy[1]), wrappers)
          sel.append(pairs[0])
          sel.append(pairs[1])
  cmd.delete(' _ nonpolar')
  cmd.delete(' _ selection')
```

```
    if len(sel) > 0:
        cmd.show _ as('dashes', name)
    elif not quiet and len(hb) != 0:
        print ' - no dehydrons were found - '
    else:
        print ' - no hydrogen bonds were found - '
############ main routine ###########
proteins = glob.glob('*.pdb')
proteins.sort()
for protein in proteins:
    cmd.load(protein)
    dehydron(selection='all', angle _ range=40, max _
distance=3.5, desolv=6.5, min _ wrappers=19, quiet=0)
    cmd.delete("all")
```

A.2 MACHINE LEARNING METHOD TO INFER STRUCTURE WRAPPING AND DEHYDRON PATTERN IN THE ABSENCE OF PROTEIN STRUCTURE: THE TWILIGHTER

When a protein has no reported structure, a dehydron prediction from protein sequence is required. This imperative becomes particularly pressing for drug designers that often face the problem of a lack of structural information on the target protein. As noted in Chapters 4 and 5, dehydron-rich regions entail a significant exposure of the protein backbone that translates into a significant propensity for native structural disorder. Thus, sequence-based predictors of disorder can provide the signal for identifying dehydrons in the missing structure. The publicly available Predictor for Native Disorder PONDR® is a tool of choice to infer dehydrons from protein sequence. PONDR may be freely accessed from the internet site with URL: http://pondr.com

PONDR provides a coarse resolution, generating a smeared-out plot resulting from prediction of disorder propensity on a sliding window along the protein sequence. We need to deconvolute the smeared-out signal to obtain individual amino acid values of disorder propensity, as described below. When using PONDR, one can only report averaged trends over sequence windows, not individual disorder propensities. This is so because the dehydron prediction with PONDR is only a statistical inference over a sliding window. The broad disorder regions ($f>0.5$) without disorder certainty usually have a fine-grained sawlike structure embedded in them with many order-disorder twilight zones rich in dehydrons. This fine structure is usually missed in a PONDR reading because the latter provides a coarse-grained picture which is adequate to predict large disorder regions (PONDR's intent) but not adequate for obtaining detailed dehydron prediction at the residue-level resolution that is necessary for the drug designer. In PONDR's broad signals, we are looking at the envelope of a sawlike structure, and in the reported envelope we are missing the dehydron-rich twilight regions.

A protocol of how to resolve PONDR signals to the level of fine graining needed to infer specific dehydrons is provided subsequently. The raw data uses PONDR to infer dehydrons but uses also structural information (templates in a training set for supervised learning) to get the level of resolution needed to generate dehydron predictions. In other words, the output from a structure-based dehydron calculator (Section A.1) is needed to infer dehydrons from PONDR. The dehydron calculator enables one to construct a training set that is used to "learn to infer dehydrons" from PONDR plots through learned resolution enhancement (LRE). Once a smeared-out (window-averaged) PONDR signal is obtained, the signal has to be resolved at the amino acid level in order to identify the twilight regions rich in dehydrons. This process of LRE is implemented by developing a computational resource, hereby named *Twilighter*, based on supervised learning technology trained with [sequence/PONDR-score/(# of dehydrons)] windows extracted from the PDB and from the PONDR plots generated for PDB-reported proteins. Once trained, the Twilighter maps the number of dehydrons onto a sequence/PONDR-score compound window. A goal is to generate the wrapping parameter $v=D/H$ (D=number of dehydrons, H=number of backbone hydrogen bonds) subsumed in the smeared-out PONDR plot.

Since dehydrons are local features and disorder predictions cannot be resolved to the level of locality required to infer the presence of individual dehydrons, a scale or resolution problem arises when PONDR is used for dehydron inference. The twilight regions corresponding to dehydrons are often smeared out as the window is slid along the sequence, and often the dehydron microstructure embedded in the PONDR signal is lost due to poor resolution.

This problem is akin to probing reaction kinetics at different temperatures and finding different orders for the chemical reaction depending on the temperature. Fine-grained features of the potential energy surface that are smeared out at high temperature become important when thermal fluctuations scale down, dramatically altering the observable kinetics of the reaction.

In our particular context of interest, a helical structure with "frayed ends," rich in dehydrons (cf. Figure 5.1(b)), will be typically "read" by PONDR as a broad disorder signal with no certainty of disorder, unless the length of the helical region is incommensurately larger than the length of PONDR sliding window. The resolution problem is solved exploiting unsupervised learning technology to interpret and deconvolute the broad signals generated by PONDR. Here we describe the supervised learning algorithm that we have aptly named Twilighter that deconvolutes PONDR-signals to identify dehydrons by learned resolution enhancement [4].

Twilighter is easily implemented, and adopts as training set the direct dehydron identification obtained using Dehydron Calculator (Section A.1) for a large proportion (typically 66%) of the proteins reported in the PDB combined with the PONDR plot output on the protein sequences for the same PDB-entries. Each PONDR signal is discretized as a 9-tuple of disorder propensities corresponding to a typical PONDR-sliding window, and the training data is represented by combining the outputs of PONDR and Dehydron Calculator into a 19-tuple: 9 entries for amino acid sequence window, 9 entries for PONDR scores, and 1 entry for number of dehydrons obtained from Dehydron Calculator. In this way, the learning machine is trained to interpret each PONDR signal from a PDB-reported protein as containing a well-determined number of dehydrons that would not be identifiable from direct reading of the PONDR plot. This training is then used to make sequence-based dehydron inferences based on 9-tuples of PONDR scores for 9-amino acid sequence windows.

The testing set for this machine is the rest (approx. 33%) of PDB-reported proteins, where we can directly contrast a PONDR-based learned dehydron prediction assigning # dehydrons to each 18-tuple (window sequence identity + the 9 respective PONDR scores) with the actual identification of dehydrons obtained using Dehydron Calculator.

A training set amounting to roughly 66% of PDB (40,719 entries) gives a prediction accuracy of 93%, with most outliers being small floppy peptides (N<35) with structure determined from NMR.

This tool is needed to extract wrapping information from sequence-based disorder prediction, since the latter is an attribute averaged over a sliding sequence window, while dehydrons pair two specific residues, thus requiring fine resolution at the single amino acid level. Thus, a learned resolution enhancer (LRE) was required to translate PONDR signals into dehydron patterns. A feedforward neural network resource can be built trained with the wrapping and PONDR-based disorder scores from

PDB-represented proteins in order to infer wrapping patterns of foldable proteins with unknown or unreported structure.

The network may be trained with information on proteins with PDB representation. The training data is represented as a vector (**s, w,** F) = (**X**, F), where **s** is the protein sequence window of length N = 9, **w** is the 9-vector of "raw" PONDR scores in the window, **X** = (**s, w**), and F is the number of residues within the window that are paired by dehydrons. The F-values for proteins in PDB (training set) are computed using Dehydron Calculator (Section A.1).

The inference is then defined by input **X** and output F = F(**X**) for a protein with unknown structure, for which **X** is determined using PONDR. Thus, learned resolution enhancement is simply an in silico lens used to get information from a smeared signal. The exact workflow of the LRE is described in Figure A.1.

The LRE uses the standard learning strategy. The idea is to generate the function F(**X**) = (# residues paired by dehydrons) from a data representation **X** = (**s, w**). For simplicity, we discretized the single amino acid score f

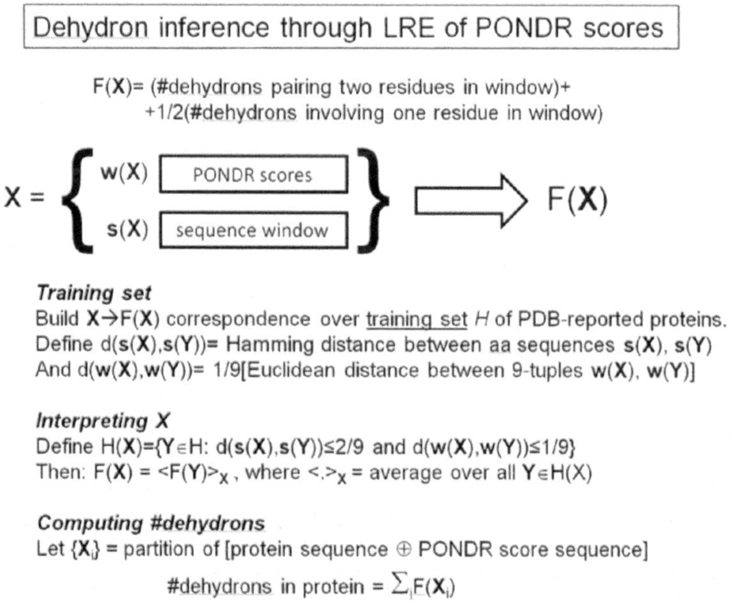

Dehydron inference through LRE of PONDR scores

F(**X**)= (#dehydrons pairing two residues in window)+
+1/2(#dehydrons involving one residue in window)

$$X = \left\{ \begin{array}{l} \text{w}(X) \quad \boxed{\text{PONDR scores}} \\ \\ \text{s}(X) \quad \boxed{\text{sequence window}} \end{array} \right\} \Longrightarrow F(X)$$

Training set
Build **X**→F(**X**) correspondence over training set *H* of PDB-reported proteins.
Define d(s(**X**),s(**Y**))= Hamming distance between aa sequences s(**X**), s(**Y**)
And d(w(**X**),w(**Y**))= 1/9[Euclidean distance between 9-tuples w(**X**), w(**Y**)]

Interpreting X
Define H(**X**)={**Y** ∈H: d(s(**X**),s(**Y**))≤2/9 and d(w(**X**),w(**Y**))≤1/9}
Then: F(**X**) = <F(**Y**)>$_X$, where <.>$_X$ = average over all **Y** ∈H(**X**)

Computing #dehydrons
Let {**X**$_i$} = partition of [protein sequence ⊕ PONDR score sequence]

#dehydrons in protein = Σ_iF(**X**$_i$)

FIGURE A.1 Workflow of the learned resolution enhancement (LRE) machine "Twilighter" used in [4] to infer dehydrons from sequence-based predictions of disorder propensity.

in three ranges: low (1/3) 0 < f < 0.35; medium (2/3): 0.35<f<0.8 ("dehydron range") and high (1): f>0.8.

We defined the training set H as the set of **X**-vectors obtained from PDB-reported proteins and their PONDR scores. We define the standard metric d in sequence space and the standard metric d in the space of ternary w-vectors. We of course know the F(**Y**) values for all **Y**s in H, because we can compute them directly from structure using the Dehydron Calculator (Section A.1).

For crude inferences, given an **X** not belonging to H, we infer the value F(**X**) by defining the set $H(\mathbf{X})$ as the set of vectors in the training set H that are "closest" to **X**. The set $H(\mathbf{X})$ is constructed using the BLAST tool (available at the US National Library of Medicine site http://blast.ncbi.nlm.nih.gov/Blast.cgi). Then, we estimate F(**X**) as the [integral part] of the average of F(**Y**) over all **Y**'s in $H(\mathbf{X})$.

To get the exact result F(**X**), we exploit a basic property of the network: For any two vectors (**X**$_1$, F$_1$) and (**X**$_2$, F$_2$) in the training set, F satisfies the strong "continuity" relation:

$$F_1 - F_2 = q\left[d_w\left(s_1, s_2\right) + \left(3/2\right) d_{PONDR}\left(w_1, w_2\right) \right], \qquad (A.1)$$

where d_w is the wrapping quasi-distance between the two sequence windows, defined as the difference in the number of carbonaceous side-chain nonpolar groups; d_{PONDR} is the compound difference in the PONDR scores (residue-by residue) in the window, and q = 0.16 is a constant. The value q = 0.16 (approx. 1/(4+2)) is obtained by noting that $\underline{1}$ new dehydron is created when the number of wrappers in the window decreases by $\underline{4}$ and $\underline{2}$ residues qualitatively raise their discretized PONDR score (1/3 ➔ 1) as defined above. Thus, F(**X**) is "continuous" in the sense that F(**X**$_2$)–F(**X**$_1$) = $q\Delta$(**X**$_2$, **X**$_1$), where

$$\Delta\left(\mathbf{X}_2, \mathbf{X}_1\right) = \left[d_w\left(s_1, s_2\right) + \left(3/2\right) d_{PONDR}\left(w_1, w_2\right) \right] \qquad (A.2)$$

Then the rigorous way to infer the # dehydrons in a window takes advantage of the "continuity" relation for the neural network output F. Thus, for **X** not in H we get:

$$F\left(\mathbf{X}\right) = F\left(\mathbf{X}^*\right) + q\Delta\left(\mathbf{X}, \mathbf{X}^*\right), \qquad (A.3)$$

where $\mathbf{X}^*$ belongs to H (hence, $F(\mathbf{X}^*)$ can be computed with certainty using Dehydron Calculator, Section A.1) and is obtained by blasting $\mathbf{X}$ in H ($\mathbf{X}^*$ realizes the minimum distance between $\mathbf{X}$ and H).

To summarize, dehydron predictions can be adequately generated from sequence-based disorder propensity inference using supervised learning technology trained with PDB-reported structures, their structure-based dehydron pattern, and their PONDR plots. The computational toolbox presented in this section is extremely useful for the design of drugs targeting proteins with unreported or unknown structure and to build pharmacoinformatics platforms, as shown in Chapters 4 and 5.

A.3 AI PLATFORM TO EMPOWER MOLECULAR DYNAMICS

Molecular dynamics (MD) remains the most valuable tool for the predictive understanding of the behavior of biological mater at the atomistic level. However, its true potential remains largely untested because relevant timescales are often beyond reach, limited portions of conformation space can get sampled, and infrequent events of direct relevance to biophysical processes are missed. A culprit for these shortcomings is the huge informational burden that is required to iterate the integration steps in order to generate an MD trajectory. To address the problem, we apply deep learning to (a) encode the dynamics into a simplified embodiment that retains only essential topological features of the vector field that steers MD integration and (b) to propagate the simplified trajectory beyond the timescales accessible to atomistic MD. We simplify the flow via an equivalence relation that identifies conformations within basins of attraction in potential energy and encodes the dynamics onto a modulo-basin "quotient space" where fast motions are averaged out. Encoding MD-generated information in quotient space enables coverage of physically meaningful timescales while unraveling the underlying dynamic hierarchy. Deep learning is exploited to propagate the coarse-grained state and to reconstruct it back at the atomistic level at a later time. As shown, the quotient-encoding-propagating-decoding scheme generates within few GPU hours protein folding pathways with experimentally verified outcomes. By contrast, MD computations covering comparable timespans would take hundreds of days on today's fastest special-purpose supercomputers. Thus, quotient space constitutes a generative model that provides hierarchical understanding of MD simulation while enabling access to realistic timescales and the capture of physically meaningful rare events.

A.3.1 Dynamical Feature Extraction for AI-Empowered Protein Folding Simulations

This section places additional demands on behalf of the reader, as it requires a certain level of familiarity with abstract mathematical thinking. The goal here is to empower molecular dynamics (MD) by placing it within an AI platform to enable the investigation of physically relevant phenomena. This empowerment requires that we introduce a simplified version of conformation space, an adiabatic version of the space, where fast motions are hierarchically averaged out within a mathematical construct named *quotient space*. The mathematical process needed to simplify the informational universe is akin to an adiabatic approximation that casts the vector field that steers MD integration as a topological simplification. Thus, the quotient space retains only the topological determinants of the flow that become operative at relevant (long) timescales. The canonical projection of the MD trajectory in quotient space may be digitally represented in a spatial tensor-like array; hence, it can be subsumed into TensorFlow, the underlying operational scheme for deep learning platforms (cf. Chapter 1). This adiabatic version of the trajectory living in quotient space may be readily propagated with enough training of the underlying neural network. In this way, the mathematical procedure of transference to quotient space places within reach the physically relevant timescales that would be otherwise inaccessible to MD computations.

Molecular dynamics (MD) computation has been widely adopted as a primary method to obtain an atomistic understanding of observable kinetics and thermodynamic properties and for inferring singular events in biological matter [1–3]. Yet, its full potential has not been realized, and its efficacy remains essentially untested. Several factors contribute to this state of affairs, especially the huge informational burden that needs to be carried over from one move to the next as the equations of motion are stepwise integrated, a process requiring full atomistic description of the starting state at each stage [4–7]. Thus, only a limited sampling of conformation space is feasible, physically relevant timescales are usually beyond reach [4–9], and rare events remain largely undiscovered [4–9]. In addition to these shortcomings, the accurate physical picture of key biomolecular processes such as protein folding or ligand-induced folding gets further befuddled by a morass of seemingly elusive systematic and conceptual errors that get amplified as the MD trajectories are propagated in time.

To overcome such problems, various procedures have been introduced, all geared at promoting the escape from basins of attraction in the potential energy surface, thereby inducing interbasin transitions that presumably improve the efficiency of conformational sampling [4–7, 9–11]. Here we describe a different construct that stems from the inherent topological architecture of the MD-steering vector field, hence expediting exploration of conformation space without resorting to ad hoc assumptions or heuristic topological models whereupon coarse-graining encoders need to be variationally optimized [12]. The simplification introduced requires an equivalence relation "~," whereby two microstates/conformations are regarded as equivalent if and only if the respective local conformations of the constitutive subunits lie in the same basins of attraction in the local potential energy surface. This "modulo basin" representation constitutes a quotient space made up of equivalence classes [13]. In generic terms, if x_i indicates a microstate of a system of N subunits, the modulo basin class (state) to which x_i belongs is denoted $\bar{x}_i$ and may be expressed as a Cartesian product of N basins designating the basin occupancies for the individual subunits: $\bar{x}_i = \prod_{n=1}^{N} B(i,n)$, where $B(i, n)$ is the basin of attraction occupied by subunit n when the system is in microstate x_i. Thus, for any two microstates x, y: $x \sim y \Leftrightarrow \bar{x} = \bar{y}$. Since the products of basins $\mathcal{B} = \prod_{n=1}^{N} B(n)$ constitute a denumerable set, the classes/states in quotient space may be conveniently designated by encoding vectors or even represented as points on a real line interval, as shown below.

Because the quotient space averages out fast motions (intrabasin equilibrations) respecting the inherent topology of the dynamical system, it represents a generative model providing a hierarchical predictive understanding of MD simulations. Evidence in support of this key point is provided in this work, as protein folding pathways with experimentally verified outcomes are generated using the quotient space construct.

To enhance the focus, from now on we consider protein folding trajectories defined by the evolution of backbone torsional coordinates $\{\phi_n, \psi_n\}_{n=1, \ldots N}$ [4–7]. Conformation space Ω thus becomes a 2N-dimensional torus, that is, the product of 2N unit circles, one of each dihedral coordinate of the protein backbone. In this context, the modulo-basin quotient space, denoted $\Omega/\sim$, is made up of Cartesian products of the so-called Ramachandran basins [13, 14], that is, the allowed low-energy regions in the potential energy surface of each residue along the protein chain of length N (Figure A.2; Chapter 2: Figures 2.7 and 2.8; Chapter 3: Figure 3.1). The topology of $\Omega/\sim$ is inherited from that of Ω, in the sense that a subset $Q \subset \Omega/\sim$ is open if and only if $\pi^{-1}Q \equiv \bigcup_{\bar{y} \in Q} \pi^{-1}\bar{y}$ is open in Ω, where $\pi : \Omega \rightarrow \Omega/\sim$ is the canonical projection that associates each micro-state with its modulo-basin class: $\pi y = \bar{y}$. A state in $\Omega/\sim$ may be labeled with a 4N-vector $(\mathbf{b}_1|\mathbf{b}_2|...|\mathbf{b}_n...)$ consisting of N binary 4-tuples $\mathbf{b}_n$ ($n = 1,2,..., N$) indicating the Ramachandran basin occupancy for each residue, so the value 1 at entry m in $\mathbf{b}_n$ ($b_{nm} = 1$) indicates that residue n occupies basin m ($m = 1, 2, 3, 4$) (Figure A.2). For example, (1000|0001) indicates the state of a dipeptide with first residue in basin 1 and second residue in basin 4. The classes/states in $\Omega/\sim$ constitute a denumerable set ordered by associating the 4N-vector with the rational number $0.b_{11}b_{12}b_{13}b_{14}b_{21}b_{22}b_{23}b_{24}...$ in decimal representation, where b_{nm} is the binary m-th entry for the n-th

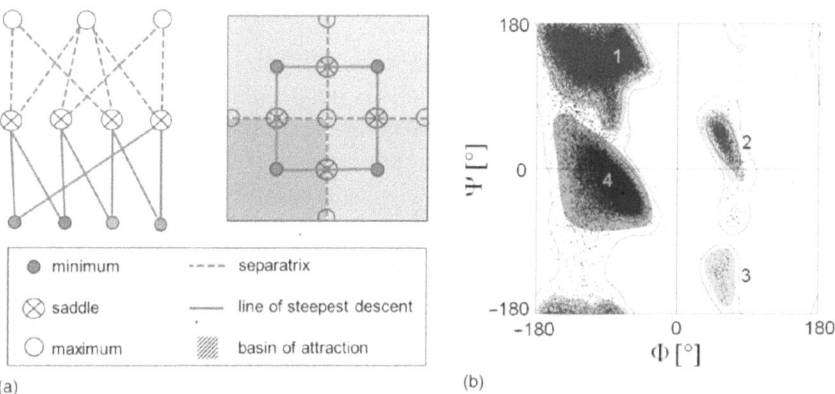

minimum   ---- separatrix

saddle   ——— line of steepest descent

maximum   //// basin of attraction

(a)   (b)

FIGURE A.2 Quotient space simplification of the vector field integrated in MD computations. (a) Topological representation of the vector field steering the backbone torsional dynamics of a generic residue in the protein chain. Opposite sides of the square are identified as per the ±180° identification of ϕ, ψ dihedrals determining the local torsional state of the backbone. Thus, the residue conformation space constitutes a 2-torus, that is, the Cartesian product of two circles. The four colored sectors morph topologically into the allowed valleys in the Ramachandran potential energy plot. Except for glycine (G), the gray sector is energetically inaccessible to other residues due to steric clash with the side chain. Other accessibility restrictions apply to geometrically constrained proline (P) and adjacent residues. The basin topology with its distribution of critical points (minima, saddles, maxima) is compatible with the underlying 2-torus. The graph on the left describes the topological organization of the basins of attraction of the critical points (minima and saddles). The basins of attraction of saddles are one-dimensional separatrices of the two-dimensional basins of minima. (b) The actual basin regions 1, 2, 3, 4 and equipotential lines in the Ramachandran plot from the collected torsional data from the MD simulations described in this work.

(Continued)

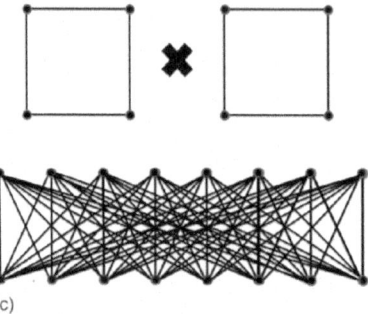

(c)

FIGURE A.2 (*Continued*) (c) Quotient space for two consecutive residues on a protein chain. For each residue, the quotient space is represented by a graph with vertices indicating two-dimensional basins, and an edge linking two vertices indicates that the respective basins are connected through a line of steepest descent crossing a saddle point and orthogonal to the separatrix at the saddle. For two residues, denoted 1 and 2, the quotient space graph is a tensor product where vertices now denote basin pairs (B_1, B_2) where basin pair (B_1, B_2) is connected with (B'_1, B'_2) if and only if B_1 is connected with B'_1 or $B_1 = B'_1$, and B_2 is connected with B'_2 or $B_2 = B'_2$. Thus, the quotient space for two residues consists of 16 vertices, where each vertex connects via one edge with 8 other vertices and connects via two adjacent edges to the 7 remaining vertices. The quotient space for two consecutive residues is shown at the bottom. An MD trajectory projected onto the quotient space becomes a walk on the graph, that is, a sequence of adjacent vertices. Reprinted from [Fernández A (2020) Rapid Research Letter: Deep Learning Unravels a Dynamic Hierarchy While Empowering Molecular Dynamics Simulations. Annalen der Physik (Berlin) 532:1900526] with permission from John Wiley and Sons.

4-tuple. By adopting this representation, the states in $\Omega/\sim$ may be plotted on a one-dimensional axis as an ordered set of rational numbers contained in the real line interval [0.00010001..., 0.10001000...].

Within this mathematical construct, we may project the MD trajectory on the quotient space and then adopt deep learning to propagate the encoded "shorthand" dynamics to simplify computations and make them more agile. The shorthand dynamics carry far lower information burden as conformations are lumped into classes that entail a considerable simplification of conformation space, while the transition timescale for each iteration is the minimal Ramachandran intrabasin equilibration time $\tau = 100$ ps [13], considerably larger than the femtosecond integration timescales used in MD. Hence, since intrabasin events are lumped together within the equivalence classes and the iteration timescale is five orders of magnitude longer, the propagated modulo-basin dynamics has a far better chance of capturing the infrequent or rare interbasin transitions, a highly sought-after goal and a significant shortcoming in MD.

A.3.2 How to Extend Molecular Dynamics Simulations

Since modulo-basin states are denumerable and the time step for the dynamics in quotient step is τ, the propagation actually indicates iterations of a transition matrix operator Γ_τ that satisfies the commutative diagram:

$$x(t) \quad \overset{\pi}{\to} \quad \overline{x(t)}$$

$$\downarrow MD \qquad \downarrow \Gamma_\tau$$

$$x(t+\tau) \overset{\pi}{\to} \overline{x(t+\tau)}$$

The diagram commutativity informs us that at any time t, running MD for a time period τ takes microstate $x(t)$ to $x(t + \tau)$, and the modulo-basin class for this destiny microstate, $\overline{x(t+\tau)}$, is identical to $\Gamma_\tau x(t)$.

To determine the transition operator Γ_τ that satisfies the diagram commutativity, we exploit a deep learning platform trained on MD data. To that effect, we break up the MD trajectory spanning a time period $[0, t_f]$ into two regions, a training portion with simulated timespan in the interval $[0, t_0]$ and an optimization portion covering the time interval $(t_0, t_f]$. With the trajectory statistics drawn for the region $[0, t_0]$, we can obtain a first approximation to Γ_τ as a maximum likelihood estimation [15]. Thus, the probability for the transition $\overline{x}_i \to \overline{x}_j$ is estimated at $\left[\Gamma_\tau\right]_{ij} = \dfrac{M_{ij}(\tau)}{M_i}$, where $M_{ij}(\tau)$ is the number of transitions $\overline{x}_i \to \overline{x}_j$ that occurred for the projected trajectory between t and $t + \tau$ within the time period $0 \leq t \leq t_0 - \tau$, and M_i is the number of times state $\overline{x}_i$ is visited within the same training time frame. Once this first estimation is at hand, the optimization of the transition operator requires deep learning as the projection of each MD-generated microstate $x(t_0 + n\tau)$ for $n = 1, \ldots, W$ and W satisfying $(t_0 + W\tau) \leq t_f < [t_0 + (W + 1)\tau]$ must be approximated as $[\Gamma_\tau]^n \pi x(t_0)$. In other words, the parametrization of the transition operator in quotient space is optimized to minimize the loss function $\mathcal{L}(\Gamma_\tau)$ [16] given by

$$\mathcal{L}(\Gamma_\tau) = W^{-1} \sum_{n=1}^{W} \left\| \pi x(t_0 + n\tau) - \left[\Gamma_\tau\right]^n \pi x(t_0) \right\|^2 \qquad \text{(A.4)}$$

We know this loss function is the correct one since $\mathcal{L}(\Gamma_\tau) = 0$ if and only if the transition operator makes the previous diagram commutative. Once the optimal $\Gamma_\tau^* = argmin\ \mathcal{L}(\Gamma_\tau)$ has been obtained from stochastic steepest descent [16], the modulo-basin projected trajectory can be propagated beyond MD-accessible timescales. At this stage, the coarse-grained modulo basin trajectory needs to be decoded back to the MD-level atomistic description. This reconstruction process also requires deep learning as we need to optimize a 1-to-1 application $\lambda : \Omega/\!\sim\ \rightarrow \Omega$ that makes the following diagram commutative in the time range $t_0 \leq t \leq t_f - \tau$:

$$x(t) \quad \overset{\pi}{\rightarrow} \quad \overline{x(t)}$$

$$\downarrow MD \qquad \downarrow \Gamma_\tau^*$$

$$x(t+\tau) \underset{\lambda}{\leftarrow} \overline{x(t+\tau)}$$

This implies that the decoder λ must be optimized to minimize the loss function:

$$\mathcal{L}(\lambda) = W^{-1} \sum_{n=1}^{W} \lVert x(t_0 + n\tau) - \lambda \left[\Gamma_\tau^* \right]^n \pi x(t_0) \rVert^2 \qquad (A.5)$$

Again, the choice of loss function is correct since $\mathcal{L}(\lambda) = 0$ if and only if the decoder makes the diagram commute. Once the decoder has been obtained as $argmin\ \mathcal{L}(\lambda)$, we can expand the MD computation as $\lambda[\Gamma_\tau^*]^n \pi x(t_0)$ for any $n \geq W$. Thus, the quotient space becomes a canonical simplification of the dynamics, and the proposed AI application is naturally tailored to access physically relevant timescales.

We have thus described a learned quotient-encoding (π) – propagating (Γ_τ^*) – decoding (λ) (QEPD) scheme that greatly simplifies and extends the MD computation, reducing significantly the informational burden at each step by learning to propagate the simplified shorthand dynamics encoded in quotient space and to decode the shorthand trajectory back to the atomistic level. In rigorous terms, the QEPD process enables to compute the MD mapping $x(0) \rightarrow x(t = n\tau)$, $\forall\ n \geq 1$ as $x(n\tau) = \lambda[\Gamma_\tau^*]^n \pi x(0)$.

.

A.3.3 AI-Generated Protein Folding Pathways

The efficacy of this QEPD is illustrated by folding a protein chain. We selected an N = 57 chain known to be capable of folding autonomously: the thermophilic variant of the B1 domain of protein G from *Streptococcus* (PDB.1GB4). The thermophile was chosen over the wild type due to higher stability of the folded structure. Using the charmm package (free version of CHARMM) [3], we first generated a 220 µs-folding trajectory within the NPT (isothermal/isobaric, T = 298 K) ensemble.

To expand MD computations reaching subsecond timespans and create a generative model of the dynamic hierarchy enshrined in the multiscale simulation, we introduce a shorthand version of the dynamics obtained by projection onto a modulo-basin quotient space. The optimization of the propagation and atomistic reconstruction of the metadynamics in quotient space requires a deep learning (DL) platform which is scripted in Python [17] with TensorFlow [18]. The DL process yields the transition matrix operator Γ_τ that propagates over time the metadynamics resulting from projection of the MD trajectory onto the quotient space $\Omega/\sim$. To incorporate the training information, a multilayer network for DL is built and exploited whereby network training involves a progressive enrichment (as defined below) of the maximum likelihood (ML) transition probability matrix $\left[\Gamma_{ML}\right]_{ij} = \dfrac{M_{ij}(\tau)}{M_i}$ for generic transition $\bar{x}_i \to \bar{x}_j$ within the time range $0 \leq t \leq t_0$. The ML transition probability matrix is generated for different MD runs I, II, ... (I denotes the distinguished run extended through AI), where the respective coarse-grained initial condition is obtained in decimal representation using random number generation of rational numbers with binary decimals in the interval [0.00010001..., 0.10001000...] representing classes in $\Omega/\sim$. The metadynamics in quotient space is Markovian as evidenced by the fact that, if state $\bar{x}_i$ is visited, the quotients $\left[\Gamma_{ML}\right]_{ij} = \dfrac{M_{ij}(\tau)}{M_i}$ are identical irrespective of the MD trajectory generated. The number of layers is contingent on the extent of transition probability coverage required to construct the matrix Γ_{ML} through progressive incorporation of entries obtained from the different MD runs (Figure A.3(a and b)). The training limits are empirically delineated by the expediency of the propagator optimization steered by the loss function. In other words, the adequacy of the training is defined vis-à-vis the efficacy

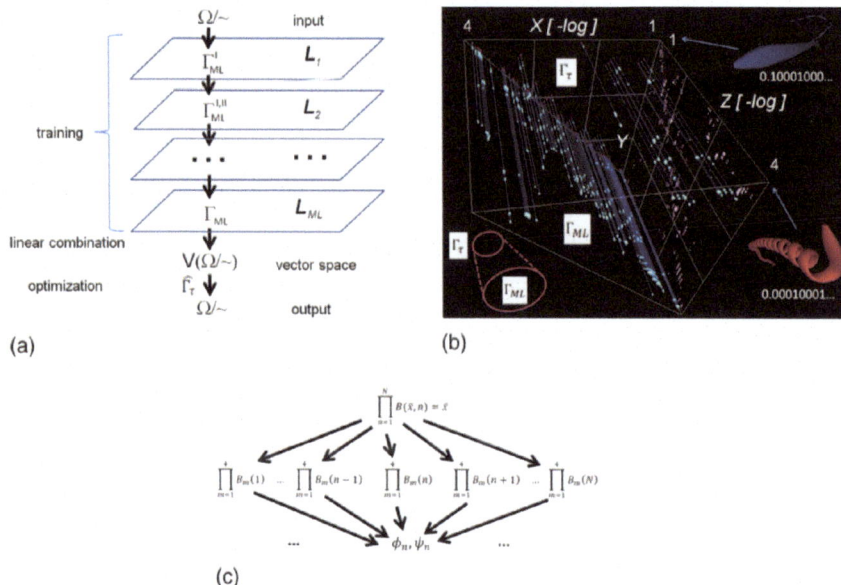

(a)

(b)

(c)

FIGURE A.3 Artificial Intelligence platform to propagate MD trajectories simplified through projection onto quotient space. (a) Scheme of multilayer deep learning network required to generate the propagation operator $\Gamma_\tau^* = \hat{\Gamma}_\tau^* \circ \Gamma_{ML}$ in quotient space $\Omega/\sim$. (b) Tensor array of transitions between modulo-basin states in $\Omega/\sim$ for the protein folding case studied. The states/classes are denoted by residue basin occupancies in decimal representation, and they are plotted on axis X, Y, Z in –log scale. The transitions dictated by optimized operator Γ_τ^* are plotted as red entries on the X-Y plane, with X and Y coordinates representing originating states and destiny states, respectively. The full set of destiny states arising from maximum likelihood transitions with probabilities ≥ 0.1 learned from 38 MD runs are plotted as light blue entries along the Z-axis. (c) Learning flow for decoding map λ used for atomistic reconstruction. For a given state $\bar{x}$ in $\Omega/\sim$ and a generic residue n along the chain, the decoded torsional coordinate region is significantly reduced beyond the constraints imposed by basin occupancy for residue n. This is so because the basin occupancy of all other residues also impose significant constraints arising from the formation of 3D-structures compatible with $\bar{x}$.

of the learning process. Clearly, multiple training runs are required for efficient learning (Figure A.4). The protein folding case illustrated in this chapter required 38 layers, and this value corresponds to minimal but sufficient training beyond which accuracy does not improve (Figure A.4).

Starting at state $x(t) = \bar{x}_i$, the destiny state of a single iteration of the modulo-basin dynamics is approximated by the fully trained network as

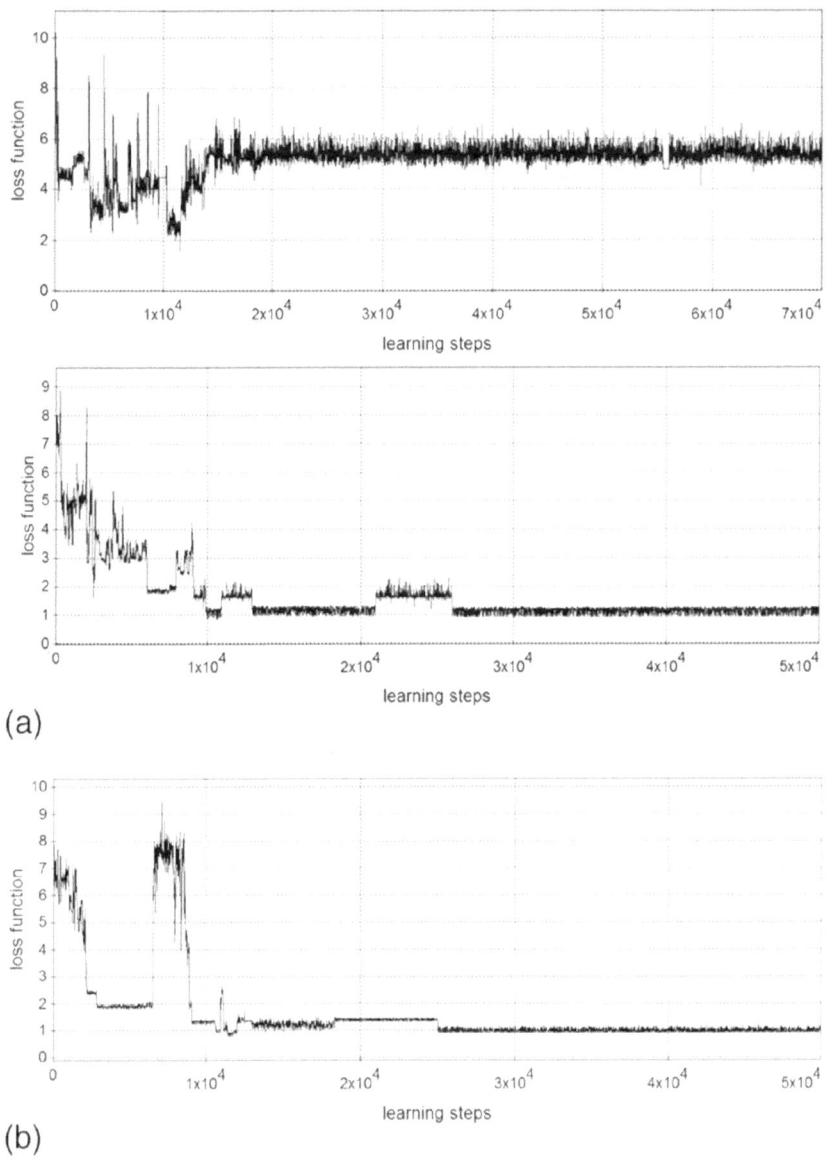

(a)

(b)

FIGURE A.4 Loss function $\mathcal{L}(\Gamma_\tau)$ for the protein folding problem. (a) The upper plot is associated with training on a single MD run, that is, the one whose extension is sought, while the plot in the lower panel displays an efficient optimization requiring 38 MD runs with randomly generated initial conditions. The steady-state value in the optimization process plateaus as the number of layers is increased beyond 38. (b) Testing accuracy computed by plotting the loss function over a testing run outside the training set of 38 MD runs. This run with randomly generated initial conditions was generated to assert the convergence to the optimization steady state.

$\overline{x(t+\tau)} \approx \left[\overline{x(t+\tau)}\right]_{ML} = \Sigma_j \left[\Gamma_{ML}\right]_{ij} \bar{x}_j,$ where the linear combination belongs to the vector space V($\Omega/\sim$) spanned by the set $\Omega/\sim$ of linearly independent vectors that constitutes the base. The linear combination state $\left[\overline{x(t+\tau)}\right]_{ML}$ serves as starting point for the parameter optimization, via minimization of $\mathcal{L}(\Gamma_\tau)$, of the retracting operator Γ_τ, that satisfies $\Gamma_\tau = \widehat{\Gamma}_\tau \circ \Gamma_{ML}$. Thus, since the maximum likelihood matrix is fixed as obtained during the training stage, the parameter optimization of Γ_τ during the period $t_0 \le t \le t_f$ boils down to identifying the optimal $\widehat{\Gamma}_\tau$ that minimizes the loss function

$$\mathcal{L}(\Gamma_\tau) = \mathcal{L}(\widehat{\Gamma}_\tau) = W^{-1} \sum_{n=1}^{W} ||\pi x(t_0 + n\tau) - \left[\widehat{\Gamma}_\tau \circ \Gamma_{ML}\right]^n \pi x(t_0)||^2 \quad (A.6)$$

Thus, the learning process is encoded in the connective weight optimization $\widehat{\Gamma}_\tau = \widehat{\Gamma}_\tau^* = argmin \, \mathcal{L}(\widehat{\Gamma}_\tau)$. The minimization of $\mathcal{L}(\widehat{\Gamma}_\tau)$ becomes a least-squares problem numerically solved with stochastic gradient descent. The set $\wp(t_0, t_f)$ of MD-generated states during the period $t_0 \le t \le t_f$ is sampled randomly at each iteration to compute the gradient with minibatches given by trajectory fragments spanned by randomly chosen time intervals of size $10^{-2} \, | \, t_f - t_0|$. To optimize the network connectivity given by $\widehat{\Gamma}_\tau$, the ADADELTA optimization protocol with learning rate 0.3 is adopted [19]. The operator $\widehat{\Gamma}_\tau$ acting on the maximum likelihood (ML) transition vector $\left[\overline{x(t+\tau)}\right]_{ML}$ has the net effect to retract the linear combination into a single destiny vector in the base $\Omega/\sim$. Figure A.3(b) displays the optimized retraction of the ML transition matrix Γ_{ML} onto the operator $\Gamma_\tau = \widehat{\Gamma}_\tau \circ \Gamma_{ML}$ for the protein folding problem described below. For clarity, only transitions with ML probability ≥ 0.1 are displayed. To generate the tensor description of the $\Gamma_{ML} \to \Gamma_\tau$ retraction, the decimal representation of residue basin occupancies was adopted enabling plotting the states in quotient space $\Omega/\sim$ on a one-dimensional axis.

A full MD trajectory encoded as the shorthand "modulo-basin" version in quotient space (Figure A.5) is shown in Chapter 3: Figures 3.3 and 3.4, with selected $t_0 = 120 \, \mu s$ as training parameter and $120 \, \mu s < t \le 220 \, \mu s$ as learning period to optimize the matrix propagator in quotient space. Using the QEPD procedure, this trajectory is further extended up to $t = 7500 \, \mu s$ (Figure A.6).

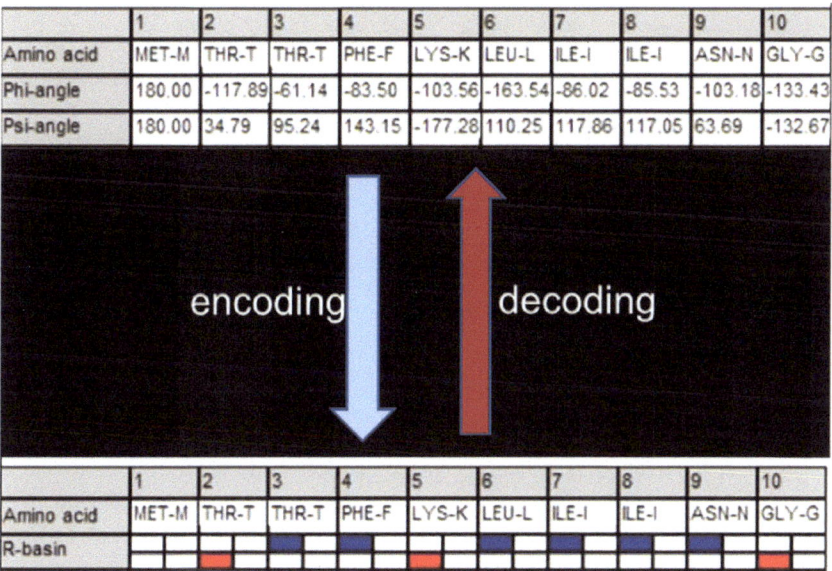

	1	2	3	4	5	6	7	8	9	10
Amino acid	MET-M	THR-T	THR-T	PHE-F	LYS-K	LEU-L	ILE-I	ILE-I	ASN-N	GLY-G
Phi-angle	180.00	-117.89	-61.14	-83.50	-103.56	-163.54	-86.02	-85.53	-103.18	-133.43
Psi-angle	180.00	34.79	95.24	143.15	-177.28	110.25	117.86	117.05	63.69	-132.67

encoding   decoding

	1	2	3	4	5	6	7	8	9	10
Amino acid	MET-M	THR-T	THR-T	PHE-F	LYS-K	LEU-L	ILE-I	ILE-I	ASN-N	GLY-G
R-basin										

FIGURE A.5 Encoding an MD trajectory in quotient space. Encoding backbone torsional information in the modulo-basin quotient space for the first ten residues in the thermophilic variant of the B1 domain of protein G from *Streptococcus* (chain length N = 57). The decoding process is the inverse operation and requires additional information on the whole trajectory encoded in quotient space (main text). A quotient-space encoded MD trajectory at 1 μs-resolution for the protein chain was presented in Chapter 3 and fully displayed in Figure A.7. A 220 μs-folding trajectory was generated within the NPT (isothermal/isobaric, T = 298 K) ensemble, selecting the timespan interval [0, t_0 = 120 μs] as training region and the interval [120 μs, 220 μs] as the learning period to optimize the parametrization of the matrix propagator in quotient space. Reprinted from [Fernández A (2020) Rapid Research Letter: Deep Learning Unravels a Dynamic Hierarchy While Empowering Molecular Dynamics Simulations. Annalen der Physik (Berlin) 532:1900526] with permission from John Wiley and Sons.

We note that the final state (Figure A.6) is significantly stable, prevailing since the time of its inception (Chapter 3). Furthermore, the decoded final stable state at t = 7500 μs, corresponding to the eigenvector associated with eigenvalue 1 of the matrix Γ_t^*, is topologically equivalent to the native crystallographic state, as evidenced by the high similarity in the contact maps, contact motifs (Figure A.6), and backbone-atom RMSD at 1.9Å. These assertions follow as the contact matrix associated with the

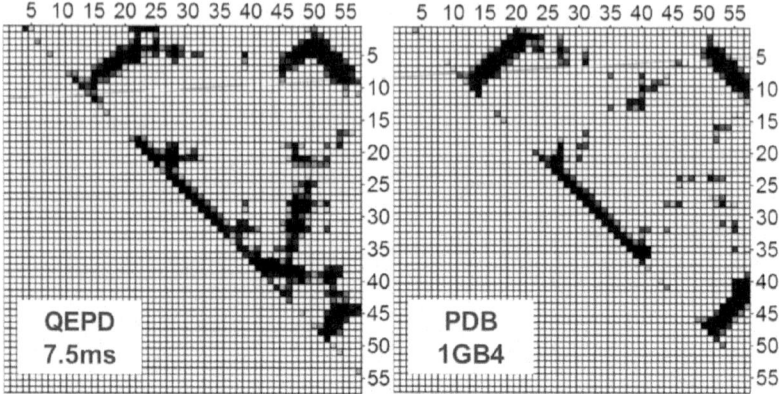

FIGURE A.6 Destiny state in propagation of an MD trajectory projected on quotient space. The MD trajectory in "modulo-basin" resolution is given in Chapter 3, Figure 3.3, at 1 μs-resolution for timespan [0, 220 μs] and has been propagated for an additional 7280 μs, spanning the overall time range [0,7500 μs] (Figure A.7). A steady state is observed in the final portion of the propagation of the projected MD trajectory (full coverage: 7500 μs), with the native-like steady state beginning to develop at around 3100 μs. The figure displays the contact map for the microstate obtained by decoding the steady state at final point t = 7500 μs of QEPD computation (left panel), equated to the eigenvector associated with eigenvalue 1 of matrix Γ_τ^*. The decoded microstate is contrasted against the topologically equivalent native state PDB.1GB4 obtained by X-ray diffraction crystallography (right panel). Reprinted from [Fernández A (2020) Rapid Research Letter: Deep Learning Unravels a Dynamic Hierarchy While Empowering Molecular Dynamics Simulations. Annalen der Physik (Berlin) 532:1900526] with permission from John Wiley and Sons.

equilibrated destiny conformation (Figure A.6) has been compared with the native fold in PDB.1GB4. An (i, j) entry in the matrix is filled in black if the minimum Euclidean distance $d_{min}(i, j)$ between all the atoms (backbone and side chain) in residues i and j is ≤ 4 Å, gray if 4 Å $< d_{min}(i, j) \leq 6$ Å, and white if $d_{min}(i, j) > 6$ Å. The two matrices reveal the same topological pattern of antiparallel and parallel β-sheets and α-helix.

We note that the atomistic trajectory obtained from the QEPD computation would take about 150 days on one of today's fastest special-purpose supercomputers covering an estimated 50 μs per day [20]. By contrast, full network training and parameter optimization took approximately 12 hours on a cluster of 12 Titan RTX GPUs, while the QEPD computation required an additional 8 hours on the same resource.

A.3.4 Molecular Dynamics Run from an AI Platform

MD has been deemed essential to gain insight into the molecular basis of thermodynamic properties and detectable events in condensed phase materials and constitutes an invaluable aid in a variety of applications [1–3]. However, exhaustive or even statistically significant sampling of conformation space becomes often prohibitively costly, and, consequently, crucial infrequent events are seldom captured in an MD computation. In the face of such limitations, we addressed the need to extend MD timespans by encoding the MD trajectory as a "shorthand" version in the so-called quotient space and then learning to propagate the encoded trajectory through the use of artificial intelligence. Quotient space is a mathematical construct that enables a simplification of the dynamics by defining an equivalence relation between microstates that are related via fast intrabasin equilibration. Thus, the class partitioning constituting quotient space is dictated by the topology of the vector field that steers the MD trajectory. Equivalence classes define how the space is dynamically related, hence significantly mitigating the information burden that needs to be carried over from one MD-integration step to the next. In this way, we introduced a scheme that makes MD computation significantly more agile, enabling coverage of physically relevant timescales while bringing meaningful insights into the dynamic hierarchy that underlies the MD computation.

In regards to extending MD timescales, coarse-graining methods represent an alternative to the quotient space projection approach. Yet, their learning efficiency requires a difficult joint variational optimization of both encoding and propagation, with an ad hoc delineation of the topology that constrains the variational search [12]. By contrast, the quotient space inherits the topology defined by class partitioning from the dynamic determinants of the system, while its inherent equivalence relation circumvents the need to optimize the encoder. Thus, the quotient space projection represents a considerably simplification and a rigorous generative model providing a hierarchical dynamics framework for the MD computation.

A.4 PROTEIN FOLDING PATHWAY GENERATED BY TRANSFORMER-ENGENDERED TOPOLOGICAL DYNAMICS

In vitro folding trajectory generated with transformer is shown in Figure A.7.

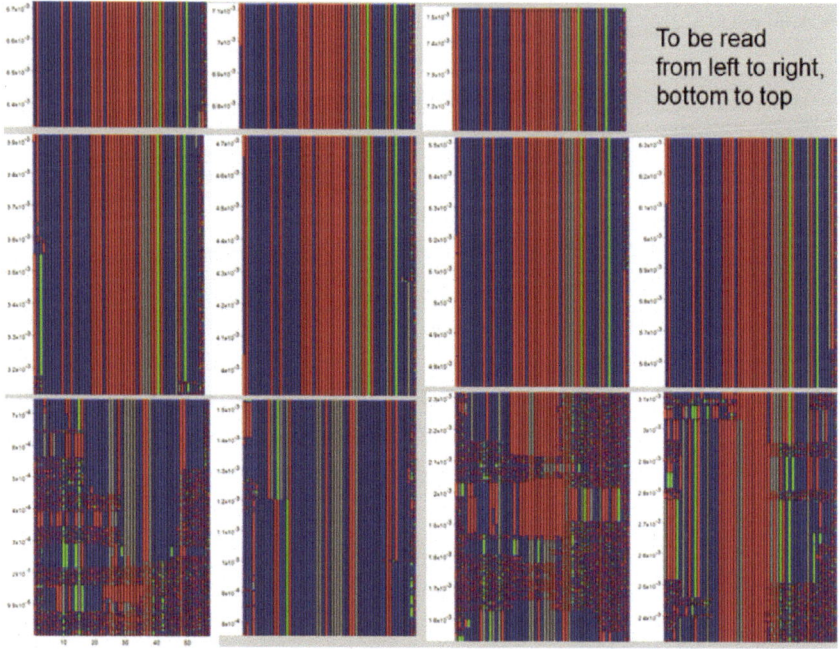

To be read
from left to right,
bottom to top

FIGURE A.7 Transformer-generated trajectory for the *in vitro* (unassisted) folding of a thermophilic variant of the B1 domain of protein G from *Streptococcus* (chain length $N = 57$, time in seconds on vertical axis, resolution at 1 μs, basin assignment for each residue along the chain on horizontal axis). The coarse-grained trajectory extends an encoded atomistic MD trajectory that covers the timespan $[0, t_f = 220 μs]$. Each state $x(t_f + n\tau)$ of the chain ($n = 1, 2, \ldots, \tau = 100$ ps) is computed as $x(t_f + n\tau) = \left[\Gamma_\tau^*\right]^n x(t_f)$.

A.5 ENDOWING THE QUANTUM MECHANICS AUTOENCODER WITH EMERGENT GRAVITY

The neural network admitting an autoencoder with emergent quantum behavior was described in Chapter 6. The emergent quantum behavior was shown to average out or thermalize the hidden variables that have been identified as components of the state **x**-vector of the network. Thus, the trainable variables conforming the **q**-vector were shown to exhibit a quantum mechanical behavior in an equilibrium regime. We assume without loss of generality that the learning process involves L separate sets of training **x**-vectors with expected values $\bar{x}^l, l = 1, 2, \ldots, L$, and these expectation vectors, together with the overall expectation vector ($l = 0$) representing the sum of all L expectation training vectors, are regarded as the hidden variables in the emergent quantum behavior of the trainable q-states. On the other hand, the nonequilibrium dynamics of the hidden variables becomes relevant on timescales much smaller than their thermalization time. This nonequilibrium dynamics is determined by the strength of the weak interactions between vector pairs $\bar{x}^v, \bar{x}^\xi, v, \xi = 0, 1, \ldots, L$ quantified by the tensor $g^i_{v\xi}$, where the dummy index i labels each neuron in the system.

To endow the hidden variables with an emergent gravity action, we first describe the nonequilibrium dynamics of the expectation vectors for the training sets. For the sake of transparency, we assume the simplest possible activation function $f = I$. Thus, for $\Delta t = \tau_< \ll \tau$, we get to first order:

$$\bar{x}^\mu_i \left(t + \tau_<\right) \approx w_{ij}\bar{x}^\mu_j \left(t\right), \quad \mu = 0, 1, \ldots, L \tag{A.7}$$

This nonequilibrium scheme yields a tangent bundle according to

$$\frac{\partial \bar{x}^\mu_i}{\partial t} \approx \tau_<^{-1}\left[w_{ij} - \delta_{ij}\right]\bar{x}^\mu_j \left(t\right) \tag{A.8}$$

And since $\bar{x}^0 = \sum_{l=1}^{L} \bar{x}^l$, we get

$$g^i_{\mu v} \frac{\partial \bar{x}^\mu_i}{\partial t} \frac{\partial \bar{x}^v_i}{\partial t} = 0 \tag{A.9}$$

where the metric tensor $g^i_{\mu v}$ describes the magnitude of the interactions between the hidden variables now cast in terms of the differential geometry of the emergent space-time.

The interactions between different training sets arise from the loss function $J_<(q)$ that holds for timescales shorter than the equilibration time for the hidden variables. This loss function becomes:

$$J_<\left(q\right) = J_<\left(w\right) = \sum_{l=0}^{L} J_<^{(l)}\left(w\right) = \sum_{l=0}^{L} \sum_{x^l \in S_l} \sum_{n=1}^{M = \left[\frac{\tau}{2\tau_<}\right]} \left|\left| x^l \left((n+1)\tau_<\right) - wx^l\left(n\tau_<\right)\right|\right|^2,$$

$$\tag{A.10}$$

where $\mathcal{S}^l \left(l = 0, \ldots, L \right)$ is the l-th training set. Thus, we define the metric tensor as

$$g^i_{\mu\nu} = \left[Argmin\left(J^{(\mu)}_< \right) - Argmin\left(J^{(\nu)}_< \right) \right]^{-1}_i \tau^{-1} \sqrt{ \int\limits_{t=0}^{\tau} \left\| \bar{x}^\mu(t) - \bar{x}^\nu(t) \right\|^2 dt }$$

(A.11)

The gravity action then becomes:

$$\mathcal{J}(w) = g^i_{\mu\nu} \left[\tau_<^{-2} \left\langle x_i^\mu G_{ij} x_j^\nu \right\rangle - \frac{\partial \bar{x}_i^\mu}{\partial t} \frac{\partial \bar{x}_i^\nu}{\partial t} \right],$$

(A.12)

with $G = G(w) = (I - w)^T (I - w)$.
Or, in Einstein's relativity terms:

$$\mathcal{J}(w) = \int dX \sqrt{-g}\, g_{\mu\nu} T^{\mu\nu},$$

(A.13)

where $g = \det(g_{\mu\nu})$ and

$$\sqrt{-g}\, T^{\mu\nu} = \left[\tau_<^{-2} \left\langle x_i^\mu G_{ij} x_j^\nu \right\rangle - \frac{\partial \bar{x}_i^\mu}{\partial t} \frac{\partial \bar{x}_i^\nu}{\partial t} \right] \prod_{l=0}^{L} \delta\left(X_i^l - \bar{x}_i^l \right)$$

(A.14)

Equations (A.10) to (A.14) define the emergent gravity of the neural network arising from the nonequilibrium dynamics of the hidden variables in the quantum mechanical autoencoder.

A.6 INCORPORATING AN EXTRA DIMENSION IN SPACE-TIME THROUGH AN AUTOENCODER OF THE STANDARD MODEL: DECODING THE HIGGS MECHANISM

A consistent incorporation of an extra spatial dimension in space-time requires that the standard 4D space-time from general relativity be regarded as the quotient (latent) manifold $\Omega = W/\sim$ endowed with the full universe of events furnished by the Standard Model. This universe is incorporated into an autoencoder, hereby named the Standard Model Autoencoder (SMAE). The ur-space (W-level) decoding of the SMAE physics is steered to incorporate an extra spatial dimension (compact or infinite, a matter of choice), as illustrated in the Epilogue (Figure E.4).

Thus, two "ur-particles" in W become equivalent vis-à-vis the relation "~" if they project onto one and the same particle in the Standard Model.

To get a glimpse of how the decoding procedure works, consider the Higgs mechanism (F_{Higgs}) in which mass is bestowed upon a particle by the field of the Higgs boson within a Langrangian dynamics scheme combining field and particle in latent space. The decoding (F) of the Higgs mechanism at the level of ur-particles is represented by the commuting diagram in the scheme below:

$$
\begin{array}{ccc}
W & \xrightarrow{\pi} & W/\sim \\
\downarrow F & & \downarrow F_{Higgs} \\
W & \xrightarrow{\pi} & W/\sim
\end{array}
$$

According to the standard model, the mass of a boson becomes the kinetic energy endowment along a vibrational mode along the curve of the zero-point potential energy in the Higgs field-particle potential energy surface. However, in the decoding of the Higgs mechanism at the level of the ur-space W, the bestowed mass of the ur-boson becomes the kinetic energy partly stored in the extra dimension. Unlike in the Kaluza-Klein models, the extra dimension, be it compact or infinite, is not treated as decoupled from the other dimensions. Hence, the problem of phase destructive interference that forces the quantization of wave-derived physical magnitudes in 4D space-time for confined particles now vanishes: an ur-particle is visible (detectable) matter when there is no out-of-phase destructive interference in the projection onto W/~, while an ur-particle becomes dark matter when the projected confined wave-particle shows destructive phase interference, and hence it is not detectable in W/~. Hence, the whole issue of quantization in the subatomic world becomes "artefactual," a consequence of the "missing" dimension in the latent manifold W/~.

REFERENCES

1. Lindorff-Larsen K, Piana S, Dror RO, Shaw DE (2011) How Fast-Folding Proteins Fold. *Science* 334:517–520.
2. Plattner N, Doerr S, De Fabritiis G, Noé F (2017) Protein-Protein Association and Binding Mechanism Resolved in Atomic Detail. *Nature Chem* 9:1005–1011.

3. Brooks BR, Brooks III CL, Mackerell AD, Nilsson L, Petrella RJ, et al. (2009) CHARMM: The Biomolecular Simulation Program. *J Comp Chem* 30:1545–1615.

4. Fernández A (1999) Coarse Graining the Soft-Mode Dynamics of a Folding Protein. *Phys Chem Chem Phys* 1:861–869.

5. Fernández A (1999) Folding a Protein by Discretizing its Backbone Torsional Dynamics. *Phys Rev E* 59:5928–5934.

6. Fernández A, Colubri A, Berry RS (2000) Topology to Geometry in Protein Folding: Beta-Lactoglobulin. *Proc Natl Acad Sci USA* 97:14062–14066.

7. Fernández A, Colubri A, Berry RS (2001) Topologies to Geometries in Protein Folding: Hierarchical and Nonhierarchical Scenarios. *J Chem Phys* 114:5871–5888.

8. Singhal N, Pande VS (2005) Error Analysis and Efficient Sampling in Markovian State Models for Molecular Dynamics. *J Chem Phys* 123:204909.

9. Zimmerman MI, Bowman GR (2015) Fast Conformational Searches by Balancing Exploration/Exploitation Trade-Offs. *J Chem Theory Comput* 11:5747–5757.

10. Buchete NV, Hummer G (2008) Coarse Master Equations for Peptide Folding Dynamics. *J Phys Chem B* 112:6057–6069.

11. Laio A, Parrinello M (2002) Escaping Free Energy Minima. *Proc Natl Acad Sci USA* 99:12562–12566.

12. Wang W, Gómez-Bombarelli R (2019) Coarse-Graining Auto-Encoders for Molecular Dynamics. *NPJ Comp Mat* 5:125.

13. Fernández A (2016) *Physics at the biomolecular interface: Fundamentals for molecular targeted therapy*. Chapter 3. Springer International Publishing, Switzerland.

14. Altis A, Nguyen PH, Hegger R, Stock G (2007) Dihedral Angle Principal Component Analysis of Molecular Dynamics Simulations. *J Chem Phys* 126:244111.

15. Rossi RJ (2018) *Mathematical statistics: An introduction to likelihood based inference*. John Wiley & Sons, New York.

16. Stanley KO, Clune J, Lehman J, Miikkulainen R (2019) Designing Neural Networks through Neuroevolution. *Nature Mach Intel* 1:24–35.

17. Ketkar N (2017) *Deep learning with python*. Apress, Berkeley, CA.

18. Rampasek L, Goldenberg A (2016) TensorFlow: Biology's Gateway to Deep Learning. *Cell Sys* 2:12–14.

19. Zeiler MD (2012) ADADELTA, an adaptive learning rate method. arXiv:1212.5701. https://arxiv.org/abs/1212.5701

20. Shaw DE, Grossman JP, Bank JA, Batson B, Adam Butts J, et al. (2014) Anton 2: Raising the bar for performance and programmability in a special-purpose molecular dynamics supercomputer. *SC '14: Proceedings of the International Conference for High Performance Computing, Networking, Storage and Analysis. IEEE ACM* 2014:41–53. DOI: 10.1109/SC.2014.9

Index